José Ferrer
Raquel Alejos

Cashew nut agro-industrial pilot plant

José Ferrer
Raquel Alejos

Cashew nut agro-industrial pilot plant

Conceptual engineering, basic engineering, detailed engineering for nut processing

ScienciaScripts

Imprint
Any brand names and product names mentioned in this book are subject to trademark, brand or patent protection and are trademarks or registered trademarks of their respective holders. The use of brand names, product names, common names, trade names, product descriptions etc. even without a particular marking in this work is in no way to be construed to mean that such names may be regarded as unrestricted in respect of trademark and brand protection legislation and could thus be used by anyone.

Cover image: www.ingimage.com

This book is a translation from the original published under ISBN 978-613-9-72884-8.

Publisher:
Sciencia Scripts
is a trademark of
Dodo Books Indian Ocean Ltd. and OmniScriptum S.R.L publishing group

120 High Road, East Finchley, London, N2 9ED, United Kingdom
Str. Armeneasca 28/1, office 1, Chisinau MD-2012, Republic of Moldova, Europe
Managing Directors: Ieva Konstantinova, Victoria Ursu
info@omniscriptum.com

Printed at: see last page
ISBN: 978-620-8-37064-0

Cashew nut agro-industrial pilot plant

José Ramón Ferrer González

Raquel Elena Alejos Pineda

SUMMARY

SUMMARY

The overall aim of implementing this project has been to improve the quality of life of cashew producers in the producing areas, through technical and economic support and the generation of direct and indirect jobs. On the other hand, it will sustainably improve the quality of life of the inhabitants of the region and neighboring areas who are currently engaged in cashew processing. An agro-industrial pilot plant is designed by developing conceptual engineering, basic engineering procedures and, finally, detailed engineering to carry out food production. The processing of cashew nuts can be summarized in four main stages: 1. receiving and weighing, drying the nuts, cleaning and washing, sorting and cooking the nuts (process and autoclave, cooling and drying the nuts); 2. cutting the nut shell, dehydrating and swelling the kernel; 3. shelling the kernels, sorting and grading and toasting and packaging. When laying out the pilot plant, a number of considerations need to be taken into account to allow for an efficient layout of equipment and space in general. In addition, this comprehensive approach will not only improve cashew nut production, but also provide valuable information on the economic viability and sustainability of the process and provide valuable information for producers and companies interested in large-scale production.

Key words: Processing, cashew nuts, sustainability, large-scale production.

INTRODUCTION

The installation of a cashew processing pilot plant is based on the fact that the crop, made up of tropical and subtropical trees and shrubs, has adapted very well in the municipalities of Rosario de Perijá and Mara, in the state of Zulia, since the existing cashew trees were not artificially introduced for commercial purposes, but propagated naturally. In these municipalities, there are hundreds of families who over the years have made a living from processing this product, whose fruit is commonly known as caujil and its seed as cashew. For generations, these families have harvested these fruits to make various products, such as sweets, jellies, cakes, nougats, and the traditional toasted cashew stands out the most.

The artisanal production process is carried out in the courtyards of houses, using a rudimentary system of heating in cans until the kernels are toasted. The nuts are separated with a bottle and packed in glass jars to be sold on the side of the road. This system has drawbacks when it comes to processing the nuts, as the people responsible for toasting them suffer burns to their hands due to the lack of suitable equipment.

In addition, the hygiene standards used in the traditional production system are not the most appropriate, as there is no control, the utensils used are rudimentary, the containers for storage are recycled and do not undergo a sterilization process.

At the same time, the demand for tropical fruits on the international market, driven by the tendency to consume nutritious foods that contain phyto-medical substances and help control diseases caused by the stress of everyday activities, among others, has led to the search for alternatives for the use of cashews.

Cashew kernels have a high nutritional value as they contain essential amino acids, fatty acids that are important for the proper functioning of the body and vitamin E, which acts as an active antioxidant. It also contains riboflavin and thiamine.

In Venezuela, the main objective of cashew cultivation continues to be the nut, while the pseudo-fruit remains underutilized. Consequently, generating products from it is important, as not using it results in 90% of the whole fruit being wasted.

The cashew pseudofruit can be transformed into a wide variety of food products, such as juices and flours, with an important content of vitamins A and C, being an adequate source of energy, proteins, minerals, essential amino acids and other valuable antioxidants, with the possibility of access to the world market. Transforming the pseudofruit into stable products by applying simple and accessible technologies would reduce post-harvest losses, diversify the use of these products and improve the diet from a nutritional point of view.

It is widely recognized that non-alcoholic beverages, especially those made from fruit, are highly consumed worldwide and are fundamental sources of vitamins and minerals for the human diet. Thus, the production of juices from pseudo-fruits could be an alternative for their diversification and utilization.

In the bakery industry, wheat flour has been the basic ingredient for making bread, and it is 100% imported. Faced with this situation, many other options have been developed, with the partial incorporation of raw materials derived from other crops. It is necessary to look for national alternatives that can replace wheat flour (which increases the cost of the final product), guaranteeing added value to national crops that have so far been underused and, at the same time, being able to fortify these products by increasing their nutritional value, as well as offering a new flavor that encourages their consumption.

The overall aim of implementing this project is to improve the quality of life of cashew producers in the producing areas, through technical and economic support and the generation of direct and indirect jobs. In addition, the aim is to sustainably improve the quality of life of the residents of the region and neighboring sectors, who are currently engaged in cashew processing.

Consequently, the specific objectives are:

- Installing a cashew processing pilot plant, which will serve as a model for the creation of future small processing units near producers' homes, facilitating access to the raw material.

- Optimize production processes and increase sales volumes.

- Increasing production yields by applying appropriate management to existing plants in the "Sabana Strip from Jalisco to 104", in the following sectors: February 2, San Juan I, San Juan, Guadalajara, La Patilla, El Carmen, La Matica I.

- Improving the marketing of cashew kernels in La Villa del Rosario, with the aim of increasing the income and livelihoods of smallholders who have plants on their properties.

- Producing a greater quantity of cashew kernels for local and regional consumption, thus contributing to food security in the region.

CHAPTER I: LITERATURE REVIEW

1.1 Agro-industrial food pilot plant

This is a small-scale facility designed to carry out experimental tests and develop food production processes at an industrial level. These plants allow concepts to be tested, processes to be adjusted and the technical and economic viability of new technologies or food products to be assessed before they are implemented on a large scale in industry.

An agro-industrial pilot plant is designed by developing conceptual engineering, basic engineering procedures and finally detailed engineering to carry out food production . For the agro-industrial processing of food products, the requirements established by Good Manufacturing Practices (GMP) and the Codex Alimentarius are taken as a basis [1, 2, 3, 4].

Good Manufacturing Practices (GMP) establish the requirements and fundamentals needed to guarantee safety at all stages of the food production process. They must consider all aspects covered by GMP, such as infrastructure (physical facilities), hygienic measures, equipment and utensils, personnel, raw materials, operations and the process cleaning system at the design stage.

1.1.1 Physical facilities

The basic design characteristics that must be met by the physical facilities of a food processing plant are established by the Codex Alimentarius, specifically in standard RTCA 67.01.33:06 [2]. They are described below:

- Factory buildings and structures must be of a size, construction and design that facilitates maintenance and sanitary operations to meet the purpose of food processing and handling, protection of the finished product and cross-contamination.

- Food industries must be designed so that they are protected from the outside environment by walls. Buildings and facilities must prevent the entry of animals, insects, rodents and/or pests or other environmental contaminants such as smoke, dust, steam or other contaminants.

- The building's environments should include a specific area for changing rooms, with appropriate furniture for storing personal items, and a specific area for eating.

- Facilities must allow for easy and adequate cleaning and inspection.

- Drawings or sketches of the physical plant must be available to locate the areas related to the production process flows.

- All building materials and installations must be such that they do not transmit undesirable substances to the food. Buildings must be of solid construction and kept in good repair.

- In the production area, wood is not allowed as one of the building materials.

As for the processing and storage areas, it is stipulated that:

1.1.2 Floors

- Floors must be made of impermeable, washable and non-slip materials that are not toxic for their intended use and must be constructed in such a way as to facilitate cleaning and disinfection.

- Floors must not have cracks or irregularities in their surface or joints.

- The joints between floors and walls should be rounded to facilitate cleaning and prevent the accumulation of materials that could lead to contamination.

- Floors must have drains and an adequate slope to allow water to drain away quickly and prevent puddles from forming.

- As appropriate, floors should be made of materials resistant to deterioration caused by contact with chemicals and machinery.

- Warehouse floors must be made of material that can withstand the weight of the stored materials and forklift traffic.

1.1.3 The walls

- External walls can be built from concrete, brick or concrete block and even prefabricated structures made from various materials.

- Internal walls must be covered with impermeable, non-absorbent, smooth, easily washable and disinfectable materials, painted in a light color and free of cracks.

- When justified by the humidity conditions during the process, the walls must be covered with a washable material up to a minimum height of 1.5 m.

- The joints between walls and floors must be concave.

1.1.4 The ceilings

- Ceilings should be constructed and finished smoothly to minimize the accumulation of dirt, condensation, mold and crusts that can contaminate food and the shedding of particles.

- False ceilings are permitted and must be smooth and easy to clean.

- The ceilings must be at least five meters high to avoid condensation of vapors in the pipes and thus the release of dirt and dust particles accumulated in the pipes and lighting fixtures.

1.1.5 Windows and doors

- Windows should be easy to clean, constructed to prevent water and pests from entering and, where appropriate, fitted with insect screens that are easy to remove and clean.

- Window openings should be sloped and of a size that prevents dust accumulation and use for storage.

- Doors should have a smooth, non-absorbent surface and be easy to clean and disinfect. They should open outwards, be well fitted to the structure and in good condition.

- Doors giving access to the outside of the processing area must be protected to prevent pests from entering.

1.1.6 Lighting

- The entire establishment must be lit by natural or artificial light in such a way as to allow tasks to be carried out and not compromise food hygiene; or by a mixture of both, ensuring a minimum intensity of 540 lux (50 candelas/ft^2) at all inspection points, 220 lux (20 candelas/ft^2) in processing rooms and 110 lux (10 candelas/ft^2) in other areas of the establishment.

- Lamps and all artificial lighting fixtures located in the raw material receiving, storage, preparation and food handling areas must be protected against breakage. Lighting must not alter colors. Electrical installations, if external, must be covered by insulated pipes or tubes, and overhead cables are not permitted in food processing areas.

1.1.7 Ventilation

- Adequate ventilation must be provided to: prevent excessive heat, allow sufficient air circulation, prevent condensation of vapors and remove contaminated air from different areas.

- The direction of air flow should never be from a contaminated area to a clean area, and ventilation openings should be protected to prevent contaminants from entering.

Once the product derived from the cashew pseudofruit has been characterized at laboratory and sensory evaluation level, the next step must be taken:

- **Supply and demand study**

The supply and demand study should be developed taking into account the results obtained from the purchase intention question in the sensory evaluation survey for marketing. Based on this information, and given the still notable absence of this sector on Venezuelan shelves, the idea of building and implementing this plant is justified.

- **Marketing channels**

Among the marketing channels, we highlight the direct distribution of the product from the production plant to warehouses, mini-markets and supermarkets, through sellers who have their own means of transport, thus minimizing the costs associated with transport and stages in the marketing chain. Social media marketing has become an efficient and high-impact advertising medium. In addition, its low cost makes the use of this communication tool even more attractive. With this in mind, we suggest the use of social networks such as Instagram, Twitter and Facebook as a means of mass promotion of the product; and as an alternative for marketing, the use of commercial websites with a high national impact, which are free for a certain period of publication, or alternatively, very low cost for the permanence of advertisements for the sale of products, such as OLX and Mercado Livre, respectively.

- **Size and location**

To establish the size of the plant, a production capacity of one ton per day (1 Ton/day) should be used as the basis for calculating the project. The location of the plant depends on certain factors, such as: location of the consumer market, sources of raw materials, availability and characteristics of the workforce, transportation facilities and suitable communication routes, availability and cost of electricity, fuel, availability of public services (water, telephone) and a space for waste disposal [1, 4, 5].

The plant must be located in an area where access to cashew plantations is easy. Due to the low water and care requirements that cashew plants need for their development, and considering the need for economic development of the indigenous communities originally from the state of Zulia, the northern zone of the state, the eastern coast, and part of the southern zone of Lake Maracaibo, where there are already plantations of this fruit, would be ideal areas for the development of the plant. One of the advantages of locating the industry in these areas would be the generation of sources of income for the local population. The consumer market, at first, would be the nearby populations, to then extend distribution to the rest of the state and even the rest of the country.

With regard to the availability of manpower, it would be necessary to offer some training lectures to the staff who will be working in the industry, because in order to guarantee the safety of the final product, it would be advisable to instruct the staff in the basic principles of good manufacturing practices, as well as the critical points in the process of making the product. Finally, a description of the production process, the distribution of equipment and an economic study of the agro-industry should be developed.

1.2 Code of hygiene practice for tree nuts

Food processing plant standards are established by the Codex Alimentarius, specifically in standard CXC 6-1972 [2, 4].

1.2.1 Scope of application

This code of practice applies expressly to almonds (*Prunus amygdalus*) and walnuts (species of the genus *Juglans*), but applies generally to all nuts produced by trees, including hazelnuts (*Corylus spp.*), pecans (*Carya illinoensis*), Brazil nuts (*Bertholletia excelsa*), cashew nuts (*Anacardium occidentale*), chestnuts (*Castanea spp.*), *macadamia* nuts *(Macadamia spp.*), etc.

This code of practice aims to provide basic hygienic requirements for orchards, for operations carried out on farms (shelling and skinning) and/or commercial operations, whether removing the shell or for nuts in shell.

This code covers all tree nuts and their derived products, including shelled, sliced, diced, ground and similar products, but does not cover products in which tree nuts are a secondary ingredient.

1.2.2 Definitions

"Vanas" are walnuts in shell that usually have an abnormally small weight as a result of intense damage of physiological origin, or caused by fungi, insects or other reasons, and which can be separated, for example, by means of a current of air.

When dealing with hygiene practices for nuts produced by trees, two basic products are recognized: the nut in the shell and the nut meat, which present specific and distinct hygienic problems. Also taken into account was the fact that a farmer dedicated to nut production can send his product to the packer either in shell or in the form of meat.

1.2.3 Raw material requirements

A) Environmental sanitation in areas where food is grown and produced

- Sanitary disposal of waste water of human and animal origin. Adequate precautions must be taken to ensure that wastewater of human and animal origin is disposed of in such a way that it does not pose a hygiene risk or health hazard, and special care must be taken to protect products from contamination by such water.

- Sanitary quality of irrigation water. The water used for irrigation must not pose any public health hazard to the consumer through the product.

- Control of plant and animal diseases and pests. When adopting measures to combat pests, treatment with chemical, biological or physical agents should only be carried out in accordance with the recommendations of the competent official authority, under the direct supervision of personnel fully familiar with the potential dangers, including the possibility of crops retaining toxic residues.

B) Harvesting and producing food under hygienic conditions

- Harvesting. Nuts produced by trees can usually be harvested by shaking the trees and collecting the nuts from the ground. Considering this method, the property should not be used for grazing cattle or housing any kind of animal. If the land has been used for these purposes, it should be plowed immediately before harvesting (with a disc harrow, rotary cultivator or other soil removal technique) to reduce the risk of fecal contamination of the nuts. When it is not possible to exclude the animals or carry out the aforementioned precautions, other measures should be adopted to protect the nuts from contamination during harvest, such as placing protective tarpaulins under the trees.

- Product equipment and containers. The equipment and containers used to package the products must not pose a health risk. Reusable packaging must be made of a material and construction that facilitates its complete cleaning, keeping it clean and in a condition that does not represent a source of contamination for the product.

- Sanitary techniques. The operations, methods and procedures followed for harvesting and production must be clean and hygienic. This includes the shelling and drying of the nuts, which are generally considered part of the harvest or operations carried out on the property.

- Peeling and drying equipment must be constructed in such a way as to allow easy cleaning and maintenance. When the cleaning operation involves the use of water, it must be potable.

- Disposal of obviously unsuitable products. Unsuitable nuts should be separated as far as possible during harvesting and production, and eliminated using an appropriate process. After shelling, it is recommended to subject all nuts to sorting to remove defective ones and to a quality inspection before sending them to subsequent operations intended for human consumption. Walnuts that show obvious fecal contamination, infestations, decomposition and other defects such as broken shells, adhering dirt, vanas, etc. to a degree that would make the product unsuitable for human consumption should not be sent for these operations.

- Protecting nuts from contamination. Appropriate precautions must be taken to protect nuts from contamination by domestic animals, insects, mites (and other arthropods), parasites, birds, chemical or microbiological contaminants or other undesirable substances during handling and storage. The type and degree of protection required will depend on the nature of the nuts and the methods used to harvest them. The nuts should be transported to a suitable storage site or to the processing site as soon as possible after harvesting and/or drying. If insect or other arthropod infestations are likely, the nuts must be subjected to fumigation or other appropriate means before storage or processing. Nuts retained for processing must be stored in closed containers, in buildings or under some form of suitable protection against domestic animals, insects, mites (and other arthropods), parasites, birds, chemical or microbiological contaminants, residues and dust. The fumigation methods and chemical products used must be approved by the competent authorities. Excessive humidity should be avoided, which favors the proliferation of fungi and the development of mycotoxins.

C) Transportation

- Means of transport. Vehicles used to transport the crop or raw product from the production area, harvesting site or storage must be suitable for their intended purpose and of a material and construction that allows for thorough cleaning. They must be cleaned and maintained in such a way that they do not represent a source of contamination for the product.

- Handling procedures. All handling procedures used must be such as to prevent contamination of the product. Special care must be taken when transporting perishable products to avoid putrefaction or deterioration. Special equipment must be used if the nature of the product or the transportation distances require it.

1.2.4 Requirements for production facilities and operations

A) Design and construction of facilities

- Location, dimensions and sanitary design. The building and surrounding area must be of such a nature that it can be kept reasonably free from undesirable odors, smoke, dust or other contaminants; it must be of sufficient size for its intended purpose, without crowding of personnel or equipment; it must be of sound construction and maintained in good condition; it must be of a construction that prevents the entry and nesting of insects, birds or vermin and it must be designed to facilitate thorough and convenient cleaning.

- Sanitary facilities and controls

a) Separation of processing operations. Areas for receiving or storing raw materials must be separate from those for preparing or packaging the final product, so as to exclude any possibility of contamination of the finished product. The areas and compartments intended for the storage, manufacture or handling of edible products must be separate and distinct from those intended for non-edible materials. The food handling area must be completely separate from the parts of the building used as staff housing.

b) Water supply. There must be an abundant supply of cold water and, where necessary, an adequate supply of hot water. The water must be potable, meeting potability standards no lower than those set out in the World Health Organization's "International Standards for Drinking Water", 1971.

c) Auxiliary water supply. When non-potable water is used - such as for firefighting - it must be transported through completely separate pipes, preferably identified by color, without any cross-connection or return siphoning with the potable water pipes.

d) Installation of plumbing and wastewater disposal. All plumbing and wastewater disposal piping (including sewage systems) must be large enough to withstand maximum loads. All connections must be watertight and equipped with suitable drains and vents. Wastewater disposal must be carried out in such a way as not to contaminate the drinking water supply. The installation of pipes and the method of waste water disposal must be approved by the competent authority.

e) Lighting and ventilation. Places must be well lit and ventilated. Special attention should be paid to exhaust fans and equipment that produce excessive heat, water vapor, noxious fumes or vapors, or contaminating aerosols. Good ventilation is essential to avoid condensation (with possible dripping of water onto the product) and the development of mold in high structures, as these molds can fall onto the food. Lamps hanging over food, at any stage of manufacture, must be of a safe type or protected to avoid food contamination in the event of breakage.

f) Toilets and services. Adequate and convenient toilets must be installed, and the areas intended for these services must have self-closing doors. Toilets must be well lit, ventilated, without direct access to the food handling area and kept in perfect hygienic conditions. Hand-washing facilities must be available in the toilet and washbasin areas, with signs for staff to wash after use.

g) Hand washing facilities. Employees must have adequate facilities for washing and drying their hands, depending on the nature of the operations. These facilities must be visible from the processing plant. Where possible, the use of disposable towels is recommended, but in any case the method of drying hands must be approved by the competent authority. Services and facilities must be kept in perfect hygienic condition at all times.

B) Equipment and utensils

Materials. All surfaces that come into contact with food must be smooth, free of cracks, fissures and must not peel; these surfaces must be non-toxic and must be resistant to food products; they must withstand repeated normal cleaning operations and must not be absorbent, unless the nature of a specific process, acceptable in other respects, requires a surface such as wood.

Sanitary design, construction and installation. Equipment and utensils must be designed and built in such a way as to prevent hygiene risks and allow for easy and thorough cleaning. Fixed equipment must be installed in such a way as to facilitate thorough and efficient cleaning.

Equipment and utensils. Equipment and utensils used for contaminating or inedible materials must be labeled to indicate their use and must not be used to handle edible products.

C) Hygienic requirements for operations

Sanitary maintenance of facilities, equipment and buildings. The building, equipment, utensils and all other accessories on the premises must be kept in good working order, clean, organized and in good sanitary condition.

During activities, waste materials must be removed frequently, using suitable waste containers. The detergents and disinfectants used must be appropriate for their purpose and used in such a way as not to pose a risk to public health.

Pest control. Effective measures must be taken to prevent insects, rodents, birds and other parasites from entering and nesting in the facilities.

No domestic animals. Dogs, cats and other domestic animals are strictly forbidden in food processing or storage areas.

Employee health. Management must inform employees that anyone with infected wounds, injuries or illnesses, especially diarrhea, must report immediately to management. Management must take the necessary measures to ensure that no one with food-borne illnesses or wounds is allowed to work in areas where they could contaminate food or food contact surfaces.

Toxic substances. Rodenticides, fumigants, insecticides and other toxic substances must be stored in locked rooms and only handled by trained personnel. Only people with full knowledge of the risks and possibility of product contamination should handle these substances, or else under direct supervision.

Personal hygiene and food handling practices

a) All food factory employees must maintain strict personal cleanliness while on duty. Their clothing, including appropriate head coverings, must be suitable and kept clean.

b) They must wash their hands whenever necessary to comply with the hygienic practices required for operations.

c) Spitting, eating, smoking and chewing gum are prohibited in food handling areas.

d) Every precaution must be taken to avoid contaminating the food or ingredients with any foreign substance.

e) Minor hand injuries should be treated and covered properly with waterproof dressings. There should be a first aid kit for these cases, to avoid food contamination.

f) Gloves used for handling food must be in perfect hygienic and clean condition, made of impermeable material, unless they are unsuitable or incompatible with the work carried out.

D) Operations and production requirements.

Handling raw materials.

a) Acceptance criteria. The factory must not accept any raw material if it is known to contain decomposing, toxic or foreign substances that cannot be eliminated until they are in acceptable concentrations by normal sorting or preparation procedures during manufacture.

Special care must be taken to avoid contamination of both walnuts in shell and walnut meat by animal or human fecal matter, and if there is any suspicion that the walnuts have been contaminated by these substances, they must be rejected for human consumption. Special precautions will be taken to reject those nuts that show signs of fungal growth, in view of the danger of containing mycotoxins.

b) Storage. Raw materials stored on factory premises must be kept in conditions that protect them from contamination and infestation, and that minimize the possibility of alteration.

c) Water. The water used to transport the raw materials into the factory must be from a source of such origin or treated in such a way that it does not pose a risk to public health, and should only be used with prior authorization from the competent official body.

Inspection and classification. Raw materials, before being introduced into the manufacturing process, or at a convenient point in the process, must undergo inspection, classification or selection, as necessary, to eliminate unsuitable materials. These operations must be carried out under clean and sanitary conditions.

Washing or other preparation. The raw material must be washed as necessary to separate soil or eliminate any other contamination. The water used in these operations must not be recirculated, unless it has been adequately treated to keep it in a condition that does not pose a danger to public health. The water used in the washing, rinsing or transportation of finished food products must be of potable quality.

Preparation and manufacturing. Preparatory operations to obtain the finished product and packaging operations must be synchronized in such a way as to achieve rapid handling of consecutive units in production, under conditions that avoid contamination, alteration, putrefaction or the development of infectious or toxicogenic microorganisms.

Packaging of the finished product

a) Materials. The materials used for packaging must be stored in hygienic conditions and must not transfer undesirable substances to the product beyond the limits acceptable to the competent official body, and must provide the product with adequate protection against contamination.

b) Technical. Packaging must be carried out in conditions that prevent contamination of the product.

Conservation of the finished product. The finished product of shelled walnuts or walnut meat must have a moisture content such that it can be kept under normal conditions without significant alteration by putrefaction, fungi or enzymatic changes. Finished products can

(a) be treated with chemical preservatives at doses approved by the Codex Committee on Food Additives, as indicated in the product standards; and

(b) be heat-treated and/or packed in hermetically sealed containers in such a way that the product remains in good condition and does not change under normal conditions.

Storage and transportation of finished products. The finished product must be stored and transported in conditions that exclude contamination with pathogenic or toxicogenic microorganisms, or their development, and that protect against alteration of the product or container.

a) All finished products must be stored in dry, clean buildings, protected against insects, mites (and other arthropods), parasites, birds, chemical or microbiological contaminants, waste and dust.

b) Optimum storage conditions:

i) To achieve optimum storage conditions, the temperature should be approximately 1°C (34°F), and the relative humidity 60 to 70 percent. In temperate countries, walnuts in shell and almonds can be stored in dry warehouses in good hygienic conditions at room temperature.

ii) When storing nut-based products in conditions where they may be infested by insects and/or mites, appropriate protection methods should be used regularly.

iii) Nut products should be stored in such a way that they can be fumigated on site, or stored in such a way that they can be transferred to specific smoke rooms (e.g. smoke chambers, steel bays, etc.). Storage in cold rooms can be used, either to prevent infestation in places where there is a high probability of insects under normal storage conditions, or to prevent insects from damaging nut products.

E) Health control program

It is advisable for each industry, in its own interest, to appoint a person, whose duties should preferably be separate from production operations, to take responsibility for cleaning the factory.

The staff under their responsibility will be permanent employees of the organization, who will be well trained in the handling of special cleaning tools, the assembly and disassembly of cleaning equipment, and the importance of contamination and the risks it entails. Critical areas, equipment and materials will be given special attention as part of a permanent sanitation program.

F) Laboratory control procedures

In addition to the checks carried out by the competent official body, it is advisable for each factory, in its own interest, to have access to a laboratory for checking the health quality of nut-based products. The extent and type of this control will vary according to the different nut products and the needs of the operation.

This control must reject all nuts that are not fit for human consumption. The analytical procedures used must follow recognized methods or standardized methods, so that their results can be easily interpreted.

1.2.5 Specifications applicable to the end product

Appropriate methods for sampling, analysis and determination must be used to meet the following specifications:

A) To the extent compatible with good manufacturing practices, the product must be free of undesirable substances.

B) Analyzed with appropriate sampling and examination methods, the product:

a) must be free of pathogenic microorganisms; and

b) should not contain, in quantities that could be toxic, any substance originating from microorganisms.

C) The product must comply with the provisions on food additives and contaminants set out in the Codex product standards, and with the maximum doses for pesticide residues recommended by the Codex Alimentarius Commission.

1.3 CODEX ALIMENTARIUS Commission and Joint FAO/WHO Program
on food standards

The Codex Alimentarius Commission implements the Joint FAO/WHO Program on Food Standards, which aims to protect consumer health and ensure fair practices in food trade. The Codex Alimentarius (from the Latin Food Law or Code) is a collection of internationally adopted and uniformly presented food standards. It also includes advisory provisions in the form of codes of practice, guidelines and other recommended measures designed to achieve the objectives of the Codex Alimentarius.

The Codex Alimentarius Commission considers that these codes of practice could be used as checklists of requirements by national food control authorities. The publication of the Codex Alimentarius aims to guide and promote the development of definitions and the establishment of requirements applicable to foods, helping to harmonize them and, consequently, facilitating international trade [4, 5, 6, 7, 8].

1.3.1 Recommended International Code of Practice Codex Alimentarius General Principles of Food Hygiene (CAC/RCP 1-1969, Rev. 4 (2003))

It is people's right to expect that the food they eat is safe and suitable for consumption. Illness and damage caused by food is at best unpleasant and at worst fatal. There are other consequences too. Outbreaks of foodborne illness can damage trade and tourism, generating economic losses, unemployment and conflict. Spoiled food causes waste and increased costs, adversely affecting trade and consumer confidence.

International food trade and international travel are increasing. The result is important socio-economic benefits, but also the spread of diseases around the world. Over the last two decades, eating habits have changed in many countries, leading to the development of new food production, preparation and distribution techniques. Effective hygiene control has therefore become essential in order to avoid the harmful consequences of diseases and food-related damage to human health and the economy. All farmers and growers, manufacturers and processors, food handlers and consumers - have a responsibility to ensure that food is safe and suitable for consumption.

These General Principles establish a solid basis for ensuring food hygiene and, where appropriate, should be used in conjunction with specific hygiene codes of practice and guidelines on microbiological criteria. The document follows the food chain from primary production to the final consumer, highlighting key hygiene controls at each stage. It recommends, wherever possible, the adoption of an approach based on the HACCP System, to increase food safety, as described in the Hazard Analysis and Critical Control Point (HACCP) System and Guidelines for its Application (Annex). The controls described in this General Principles document are internationally recognized as essential to ensure that food is safe and fit for consumption.

The General Principles are aimed at governments, industries (including individual primary producers, manufacturers, processors, food service operators and retailers), as well as consumers.

1.3.1.1 Codex general principles on food hygiene [4, 5, 6, 7, 8]

- identify the fundamental principles of food hygiene applicable throughout the food chain (from primary production to the final consumer), to ensure that food is safe and suitable for human consumption;

- recommend the application of an approach based on the HACCP system as a means of increasing food safety;

- indicate how to implement these principles; and

- to provide guidance for the development of specific codes, necessary for sectors of the food chain, processes and products, in order to extend specific hygiene requirements.

1.3.1.2 Hazard Analysis and Critical Control Point (HACCP) system and guide for its application. Annex to CAC/RCP 1-1969, Rev. 4 (2003) [5, 9, 10, 11].

The first section of this document sets out the principles of the Hazard Analysis and Critical Control Point (HACCP) system adopted by the Codex Alimentarius Commission (CCA). In the second section, general guidelines are provided for the application of the system, recognizing that the details of application may vary depending on the circumstances of food processing.
The HACCP system, which has scientific foundations and is systematic in nature, makes it possible to identify specific hazards and measures to control them, with the aim of guaranteeing food safety.

HACCP is a tool for assessing hazards and establishing control systems focused on prevention rather than analyzing the end product.

Every HACCP system can be adapted to changes, such as updates to equipment design, processing procedures or technological developments.

The HACCP system can be applied throughout the food chain, from primary production to final consumption, and its application must be based on scientific evidence of risks to human health. In addition to improving food safety, the application of the HACCP system can provide other important benefits, such as facilitating inspection by regulatory authorities and promoting international trade by increasing confidence in food safety.

Successful application of the HACCP system requires the commitment and active participation of the management and staff involved. It also requires a multidisciplinary approach that should include, where appropriate, expertise in agronomy, veterinary medicine, production, microbiology, medicine, public health, food technology, environmental health, chemistry and engineering, depending on the particular subject. The application of the HACCP system is compatible with the application of quality management systems, such as the ISO 9000 series, and HACCP is preferred among the list of food safety management systems.

Although the application of the HACCP system dealt with in the document is aimed at food safety, this concept can be applied to other aspects of food quality.

1.3.1.3 Definitions

Corrective action. Any measure to be adopted when the results of monitoring the Critical Control Points (CCP) indicate a loss of process control.

Hazard analysis. The process of gathering and evaluating information õ about hazards and the conditions that determine their presence, in order to decide which are significant for food safety and should therefore be dealt with in the HACCP plan.

Control. Adopt all the necessary measures to guarantee and maintain compliance with the criteria established in the HACCP plan.

Control. Condition achieved by correctly complying with procedures and meeting established criteria.

Deviation. Failure to meet the critical limit.

Stage. Point, procedure, operation or stage in the food chain, from primary production to final consumption, including raw materials.

Flowchart. A systematic representation of the sequence of stages or operations used in the production or manufacture of a particular food product.

HACCP. A system for identifying, evaluating and controlling hazards that are significant for food safety.

Critical limit. Criterion that separates what is acceptable from what is not.

Control measure. Any measure and activity used to prevent or eliminate a food safety hazard or reduce it to an acceptable level.

Monitoring. The act of conducting a planned sequence of observations or measurements of control parameters to assess whether a CCP is under control.

HACCP Plan. A document prepared in accordance with the principles of the HACCP System to guarantee the control of hazards important to food safety in a given segment of the food chain.

Hazard. Biological, chemical or physical agent present in the food, or condition presented by the food, which may cause adverse health effects.

Critical Control Point (CCP). Stage at which an essential control can be applied to prevent or eliminate a food safety hazard or reduce it to an acceptable level.

Validation. Verification that the elements of the HACCP plan are effective.

Verification. Application of methods, procedures, analyses and other evaluations, as well as monitoring to determine compliance with the HACCP plan.

1.3.1.4 Principles of the HACCP System

The HACCP system consists of seven principles [4, 5, 6, 7, 8]:

PRINCIPLE 1

Carry out a hazard analysis.

PRINCIPLE 2

Determine the CCPs.

PRINCIPLE 3

Establish the critical limit(s).

PRINCIPLE 4

Establish a system to monitor the control of CCPs.

PRINCIPLE 5

Establish the corrective action to be taken when monitoring indicates that a given CCP is not under control.

PRINCIPLE 6

Establish verification procedures to confirm that the HACCP system is working effectively.

PRINCIPLE 7

Establish a system for documenting all procedures and records appropriate to these principles and their application.

1.3.1.5 Guidelines for applying the HACCP system

Before the HACCP system can be applied in any sector of the food chain, it is necessary for the sector to have implemented the programs considered to be prerequisites, such as Good Hygiene Practices, in accordance with the Codex General Principles of Food Hygiene, the relevant Codex Codes of Practice and the appropriate food safety requirements. These programs, considered necessary for the HAPPC system, including training programs, must be well established and fully functional, and must be verified in order to facilitate the effective application and implementation of the HACCP system.

In all types of food companies, effective implementation of the HACCP system requires commitment and awareness at management level. This effectiveness will also depend on knowledge of the HACCP system and adequate technical skills on the part of the management and staff involved.

During the identification and evaluation of hazards, and the subsequent planning and implementation of the HACCP system, consideration should be given to the impact of raw materials, ingredients, manufacturing practices, the role of manufacturing processes in controlling hazards, the likely end use of the product, vulnerable consumer categories and epidemiological evidence relating to food safety.

The main purpose of the HACCP system is to control CCPs. If a hazard is identified that needs to be controlled, but no CCP has been found, the possibility of reformulating the process step should be considered.

The HACCP system must be applied separately to each specific operation. It may happen that the CCPs identified in a given example of the Codex Code of Hygienic Practices, when applied in a specific situation, are not the only ones identified or behave differently.

When there are any changes to the product, the process or any stage, the application of the HACCP system should be reviewed and the necessary adjustments made.

Applying the principles of the HACCP system should be the responsibility of each company. However, governments and the productive sector recognize that there may be obstacles that prevent the effective application of this system by each company. This is particularly relevant for small and less developed companies.

Although it is recognized that, when implementing the HACCP system, flexibility appropriate to the company is important, the seven principles on which the system is based must be applied. This flexibility must take into account the nature of the activity and the size of the company, including human and financial resources, infrastructure, procedures, knowledge and practical limitations.

Smaller or less developed companies do not always have the resources and expertise to develop and implement an effective HACCP plan. In such cases, expert advice should be obtained from other sources, which may include trade and industry associations, independent experts and regulatory authorities. Scientific literature on the HACCP system and, in particular, guides drawn up specifically for a sector can be useful. A HACCP system guide drawn up by experts, referring to the process or type of operation, can be a useful tool for companies during the planning and implementation of the HACCP plan.

When companies use HACCP guides drawn up by experts, it is essential that they are specific to the foods and/or processes under consideration.

In the FAO/WHO document (currently being drafted), "Obstacles to the Application of HACCP, Particularly in Small and Less Developed Businesses, and Approaches to Overcome them", you will find detailed information on the obstacles to implementing the system, particularly in small and less developed businesses, and recommendations for overcoming them.

However, the effectiveness of any HACCP system will depend on adequate knowledge and skills on the part of managers and employees, and therefore requires constant training for professionals at all levels.

1. Forming the HACCP team

The food company must ensure that product-specific knowledge and expertise are available for the effective development of a HACCP plan. The ideal way to achieve this is by forming a multidisciplinary team.

When it is not possible to have such expertise in-house, specialist advice can be obtained from other sources, such as trade and industry associations, independent experts, regulatory authorities, scientific literature and recommendations for applying the HACCP system (in particular HACCP implementation guides for specific sectors). It is likely that a suitably trained employee who has access to these guides will be able to implement the HACCP system in the company. The scope of the HACCP plan should be determined and describe which segment of the food chain is involved and the hazard classes to be addressed (e.g. does it cover all hazard classes or just
selected classes).

2. Product description

A complete description of the product should be drawn up, including relevant safety information, such as composition, physicochemical structure (including Aw, pH, etc.), microbiocidal or microbiostatic treatments (heat treatment, freezing, brining, smoking, etc.), packaging, durability and storage conditions and distribution system ção . In companies that handle multiple products, such as food service companies, grouping together products with similar characteristics or processing stages can be effective for drawing up a HACCP plan.

3. Determining intended use

The intended use of the product should be based on the expected uses of the product by the user or end consumer. In certain cases, vulnerable groups should be identified, such as those who are fed in institutions.

4. Drawing up the Flowchart

The flowchart must be drawn up by the HACCP team (see also paragraph 1). The flowchart must cover all the stages of the operation relating to a given product. The same flowchart can be used for several products as long as their manufacture involves similar processing steps. When applying the HACCP system to a particular operation, the steps before and after the specified operation must be taken into account.

5. Confirmation of the Flowchart on site

Measures must be taken to confirm consistency between the flowchart and processing throughout the stages and moments of the operation, revising the flowchart if necessary. Confirmation of the flowchart must be the responsibility of person(s) with sufficient knowledge of the processing steps.

CHAPTER II: DESIGN AND CONFIGURATION OF THE PILOT PLANT

2.1 Product Description and Characteristics

Common or vulgar name: Anacardo, Marañón, Merey, Alcayoiba, Jacote marañón, Acacauba, Acajú, Acayoba, Cacahuil, Cajú, Cajuil, Caracolí, Casoi, Casoy, Casú, Caují, Caujil, Cayutero, Locote, Merei, Noz de caoba, Noz cajul, Oacajú, Pajuil, Pangí, Panjí, Paují, Paujil.
Scientific or Latin name: Anacardium occidentale.
Name in Spanish: Anacardo, noz de marañón, merey. In other languages, it is known as:
Portuguese: caju, cajueiro, pé de caju, castanha de caju, aca de caju.
French: cajou, acajou, ancardier, noix de cajou, pomme de cajou, amande de cajou.
English: cashew, cashew tree, cashew nut, cashew apple, cashew kernel.
Hindi: cadju
Sinhalese: cadju
Italian: anacardio, noce d'anacardio, mandorla d'anacardio.
Dutch: acajou, kashu.
German: acajuban, kashunuss.
Swahili: mkanju, korosho.
Somali: bibbo, bibs.
Indonesian: jambu mente, jambu mete.

2.1.1 Trunk

The merey, Anacardium occidentale L., is a crop native to northeastern Brazil with excellent medicinal and nutritional properties. It is characterized by being a tree with a developed appearance, with an approximate height of between 5 and 7 metres, evergreen, and its trunk branches at a very low height. The trunk exudes a resin and the bark releases an acrid oil. The lifespan of a tree is around 30 years and it starts producing fruit from the third year.

2.1.2 Leaves and flowers

The leaves are simple and alternate, clustered at the apical ends of the branches, usually red when young and green when mature. They are oval to rounded in shape, 9 to 15 cm long and 3 to 10 cm wide, rounded at the end, acute or obtuse at the base. They are numerous, small and irregular.

2.1.3 Fruit

The merey nut is attached to a swollen, almost spongy, juicy, fibrous and edible stalk. When ripe, it is 8 cm wide, red and yellow in color and has a pleasant, acidic and astringent taste.

The fruit is made up of two parts: the pseudofruit and the real fruit, which is the nut, located on the outside of the pseudofruit and adjacent to it. It can be classified according to its nature as perishable or non-perishable.

When it comes to the nut, it is classified as durable, and when it comes to the pseudofruit, it is classified as non-durable.

The nut is the most valuable part of the marañón. It is made up of 49-50% shell, 23% oils, 24% kernel and between 3% and 4% moisture. The kernel, which is the edible part, is very popular as a snack, either on its own or mixed with other seeds such as peanuts, macadamias and almonds.

The pseudofruit is the result of the development of the peduncle into a fleshy structure characteristic of this plant, which develops and ripens after the nut. It is very juicy, aromatic and has a high nutritional value due to its high vitamin C content (262 mg/100 ml of juice), which is 4 to 5 times higher than the content of oranges and other citrus fruits.

It also contains calcium, riboflavin, iron, phosphorus and protein, making it an important natural source of vitamins and minerals for the human diet.

The real fruit is the nut, located on the outside of the pseudofruit and adjacent to it. It is green in color, turning grey when ripe, kidney-shaped, hard and dry, about 2 to 5 cm long, where the seed is housed.

The pericarp of the walnut, specifically the mesocarp, contains an extremely caustic oil with a dark brown color and pungent taste, called cardol, composed of 55%-64% oleic acid ($C18H34O2$) and 7%-20% linoleic acid.

The merey kernel makes up around a third of the fruit's weight, and its analysis indicates a content of 55%-60% oil, 15%-20% protein and 5% carbohydrates (starch and sugar).

Almonds are rich in calories, proteins, fats, phosphorus and thiamine. The fat content of 100 grams of seeds is 45.6 grams.

In addition, almonds have a high nutritional value, as they contain seven of the eight amino acids that are essential for the maintenance of an adult and nine of the ten that play a role in the growth of children. The amino acids found in greatest quantity are glutamic acid, arginine and aspartic acid. On the other hand, it contains fatty acids, important for the proper functioning of the body, and vitamin E, which acts as an active antioxidant, preventing oxidation, as well as concentrations of riboflavin and thiamine.

The seed is in great demand worldwide due to its nutritional properties and because it has a very pleasant taste.

Nutritional content: a quarter of a cup of marañón nuts provides:

- 37.4% of the recommended daily value of monounsaturated fat.
- 38.0% of the recommended daily value of copper: Copper plays an important role in a wide range of physiological processes, such as iron utilization, free radical elimination, bone and connective tissue

development, melanin production, energy production and antioxidant defenses.

- 22.3% of the recommended daily value of magnesium: Guarantees bone health, helps reduce blood pressure, prevents heart attacks, promotes normal sleep patterns in menopausal women, reduces the severity of asthma.
- Contains less fat than most nuts.
- It contains mainly monounsaturated fatty acids, 75% of which is oleic acid, the same found in olive oil.
- oleic acid found in marañón nuts promotes good heart health.

Nutritional intake and advantages of consumption (see Table 2.1) [14, 15]:

- Fats: The fats they contain are cardio-healthy and, according to different studies, daily or continuous consumption of a small portion (25 g) reduces levels of total cholesterol and LDL (bad) cholesterol and increases levels of HDL (good) cholesterol. This whole process is mainly due to the fact that oilseeds are very rich in oleic acid and phytosterols. The latter play a very important role in our physiology, reducing the absorption of cholesterol in the small intestine.
- Proteins: In terms of protein intake, the content of an amino acid called arginine stands out (especially in walnuts). This amino acid also makes nuts a cardio-healthy food, as arginine plays a role in the production of nitric oxide, which reduces platelet adhesion and aggregation on the vascular endothelium and also acts as a vasodilator.
- Vitamins: It's important to note that the B vitamins include folic acid, which reduces the progression of the atherosclerotic process. Oilseeds are also natural antioxidants, as they protect us against the destructive action of free radicals. This antioxidant power is conferred by the high content of Vitamin E, or tocopherol. Tocopherols protect low-density lipoproteins (LDL) from being altered by free radicals.

- Minerals: Because they are low in sodium, oilseeds are also suitable for regulating blood pressure levels. They are also excellent sources of iron, magnesium, potassium, phosphorus, calcium and zinc.
- Fiber: Its fiber content helps to increase the feeling of satiety, regulate the rhythm of intestinal transit and reduce the absorption of glucose, fats and cholesterol in the body.

Table 2.1 Nutrients in oilseeds per 100 grams

Dried Fruit	Heat (kcal)	Proteins	Fats	Carbohydrates	Fibers
Almonds	599	19	54	9.3	10
Cashew nuts	564	17.2	42	29	
Hazelnut	643	13	61	10.6	7.4
Peanuts	571	26	48	8.6	7.1
Chestnut	196	3.4	1.9	41.2	1.0
Walnut	666	15	62	12.1	4.6
Pine nuts	674	13	60	20.5	1.0
Pistachio	598	20.8	51.6	12.5	6.5
Sunflower seeds	582	27	49	8.3	6.3

There are many advantages to dried fruit when it comes to eating a healthy diet, controlling the amount you consume each day and being a healthy and nutritious snack.

.

2. 1.4 Crop varieties and production areas

Countries like Brazil have extensive knowledge of the crop, both from an agronomic point of view and in terms of physical and chemical quality, which is why, through genetic improvement, they have obtained clones within the different types of cashew nuts that are grown in this region [16]. This has allowed them to classify them according to the size of the nut, making it the second largest exporter of cashew nuts, which generates a source of foreign currency for the country's economy. Similarly, the processing of cashew pseudo-fruit to extract juice and make soft drinks, liqueurs and other products with high yields has been technologized [17].

Brazilian research into this crop has led to progress in the genetic improvement of the early dwarf cashew tree at the Pacajus Experimental Station. These plants are characterized by their small size, high production and require improved management techniques [16, 18].

In Venezuela, there are no specific varieties as a result of seed propagation, but only two defined types, the scarlet red cashew and the yellow cashew. It is rarely cultivated, but is found in semi-wild form. Similarly, there are currently no improved varieties of the common type of cashew [1].

In Colombia, the existing varieties are the product of natural crossbreeding. In general terms, the color and size of the pulp, the fleshy part or "apple", are very different. The main varieties are: the giant Magdalena, yellow and red; the long Nazaré and the small Meta [1].

In Nicaragua, two types of fruit are currently known, red and yellow apples, the latter being less astringent than the red ones. Table 2.2 shows the areas of cashew cultivation in the world [1].

Table 2.2

International cashew growing areas (Anacardium occidentale L.)

Continent	Geographical location

Americas	United States (Florida), Mexico, Cuba, Haiti, Jamaica, Guatemala, Antilles, Salvador, Trinidad, Venezuela, Colombia, Peru, Brazil, Panama.
Africa	Senegal, Mali, Guinea, Ivory Coast, Ghana, Dahomey, Nigeria, Kenya, Congo, Tanzania, Angola, Mozambique, Madagascar, South Africa.
Asia	India, Vietnam, Ceylon, Indochina, Philippines, Malaysia, Indonesia.
Oceania	Hawaii, Tahiti, Australia

Table 2.3 shows the chemical composition of cashew nuts, as presented by various authors [19, 20, 21, 22, 23, 24].

Table 2.3
Chemical and nutritional composition of cashew pseudofruit

Composition	**Quantity**
Caloric intake (kJ)	125,50
Moisture (g)	88,50
Protein (g)	0,90
Total fat (g)	0,10
Carbohydrates (g)	7,70
Crude fiber (g)	2,50
Ash (g)	0,30
Calcium (mg)	14,00
Phosphorus (mg)	24,00
Iron (mg)	0,40
Sodium (mg)	16,00
Potassium (mg)	565,00
Magnesium (mg)	260,00
Zinc (mg)	5,60
Copper (mg)	2,22
Manganese (mg)	0,83

Composition	Quantity
Vitamin A* (IU)	50,00
Vitamin A* (ER)	5,00
Vitamin B1 (thiamine) (mg)	0,03
Vitamin B2 (Riboflavin) (mg)	111,50
Vitamin B3 (niacin) (mg)	0,40
Vitamin C (mg)	200,00
Total extractable polyphenols (mg)	168,25

* IU: The International Unit used to measure the biological activity of many vitamins, hormones, enzymes and medicines.
*ER: It's a unit called retinol equivalents.

The cashew nut contains a high proportion of monounsaturated fatty acids, folic acid, vitamins B1, B2 and also magnesium, calcium and potassium, which benefit the nervous system. It is rich in protein, with a large amount of arginine, and contains a good proportion of plant sterols which help to reduce the absorption of cholesterol.

2.2 Uses and Users

Cashew nuts are a crop that has many different uses, depending on which part of the plant is used. Brazil is the country that makes the most efficient use of cashew nuts, and there is a highly specialized industry for making substantial use of the plant.

2.2.1 Uses

The pseudo-fruit is used to make soft drinks, ice cream, jams, preserves, jellies, dried cashew nuts, dried cashews, wine, vinegar and juices. In India, people make brandy from the juice, and it can also be eaten as fresh fruit. Despite having great potential, this part of the sector only processes 6% of total current production, as there is a guaranteed sale only for the seeds, which are in greater demand worldwide due to their nutritional properties, being highly recommended in the diet and relatively durable [24].

Cashew nuts are the main product and constitute the real fruit, made up of the shell and the kernel, and are cultivated for their nutritional and commercial value. It is estimated that 60% of cashew nuts are consumed in the form of snacks, and the remaining 40% in confectionery or bakery [24, 25, 26].

- Culinary

- ✓ It is used in confectionery to make sweets and mixed with chocolate in the baking industry.
- ✓ Cashew butter is made from the ground kernels.
- ✓ Cashew kernels are in demand for direct consumption, either naturally or after being roasted or fried.
- ✓ Almonds are used to make a very tasty type of nougat, and
- ✓ with ground seeds, marzipan, which is also a nougat, made with milk, sugar and broken cashews.

- Medicine [24, 27]

- ✓ Nutritious (seeds). Aphrodisiac. Intellectual stimulant. Healing (external use). Antiulcer (external use). Sexual inappetence (seeds). Lack of memory (seeds). Dementia (bark). Memory loss (bark). Nervous weakness (bark). Swelling of the legs (bark). Calluses, corns and warts (external use, reddish liquid between the two cuticles and pulp juice). Chronic ulcers, eczema and psoriasis (external use, reddish liquid between the two cuticles and pulp juice).

- ✓ Cashew apple juice, without extracting the tannin, is prescribed as a remedy for sore throats and chronic dysentery in Cuba and Brazil. Fresh or distilled, it is a powerful diuretic and is said to have sweating properties.

- ✓ Oil from the bark is used to treat leprosy, cancerous ulcers, elephantiasis, psoriasis, herpes, warts, granulomas and cracks in the feet.

- ✓ Anacardic acid has shown some activity against "Walter 256 carcinoma" and is believed to be effective against amoebas, gingivitis, ulcers, skin problems and malaria.

- ✓ oil was originally used to protect poles and floors from insects such as termites.

- Industry [24, 25, 26, 27, 28, 29, 30]

- ✓ The oil from the bark of the chestnut, called cardol, is irritating and caustic and is the cheapest source of phenols. It contains 90% anacardic acid and ortho-hydroxybenzoic acid with a natural unsaturated acid chain and 10% cardol.

- ✓ The resin obtained by condensing the bark oil with formaldehyde and linseed oil or tung oil is used to make resins, varnishes and enamels for ovens.

- ✓ It is also used in industry to manufacture plastics and resins, insecticides, printing inks and for printing cotton fabrics, pharmaceutical products, gums and varnishes.

- ✓ The almond is used industrially to make cosmetics, resins, varnishes and dyes.

2.2.2 Users

The users of cashew kernels will be all the companies dedicated to making sweets and treats at the various points of sale established in Villa del Rosario, those in the city of Maracaibo and the general public.

2.3 Demand Analysis

In order to determine and measure the forces that affect the market's requirements for cashews, as well as determine their possible participation in the commercial sphere, demand is studied.

Industrial consumption of cashew kernels in Venezuela is limited to imported product, mostly from Brazil, the United States and a small part from China.

For the purposes of this study, the nut that constitutes the real fruit will be considered, from which the kernel will be obtained through processing. However, it is suggested that the pseudo-fruit be marketed as an input for processing plants [24, 31].

2.3.1 Apparent national and imported cashew consumption

One way of determining demand is through Apparent National Consumption (ANC), which is the quantity of a given good or service that the market requires. It is established on the basis of the historical series [15, 32].

In this project, in order to know the historical demand, the C.N.A. formula was used, which shows the quantity of the good in question; in this study, it refers to the kilograms of cashew that the market consumed in the period analyzed, from 2002 to 2010, and is expressed as follows:

C.N.A. = P.N. (National Production) + I (Imports) - E (Exports)

Given the characteristics of the product, it has a demand for a socially and nationally necessary good, since it is required for development and growth, and is related to feeding people.

2.3.2 Volume and value of national production

According to official data provided by the Ministry of Agriculture and Lands, Statistics Directorate, 1987-2003 statistical yearbooks, UEMPPAT, the volume and value of cashew production in Venezuela was not artificially introduced for commercial purposes, but was propagated naturally.

Productivity is expressed in kilograms (fruit) per hectare, considering that the plants are over 20 years old. The yield is 160 kg per plant, with a spacing of 10 x 10 meters, i.e. 100 plants per hectare [15, 33].

Table 2.4

Volume and value of national cashew production in 2022

State of Venezuela	Harvested area (ha)	Production (t)	Yield (kg/ha)	Total cashew nuts (t)
Amazonas	30	390,00	13.000,00	98,00
Anzoategui	85	800,00	9.412,00	800,00
Bolivar	3.600	4.307,00	1.196,00	1.077,00
Guaric	932	1.118,00	1.200,00	280,00
Monagas	1.219	1.463,00	1.200,00	366,00
Zulia	90	99,00	1.100,00	25,00
Total	5.956	8.177,00	27.108,00	2.646,00

At a national level, as can be seen in Table 2.4, the Venezuelan states with the highest cashew (or cashew nut) production in the 2022 analysis period are shown. The most important cashew areas are found mainly in the wild in the north of Bolívar state, with an average of 1,077 tons of nuts/year; followed by Monagas, which has 366 tons of nuts/year; and Guárico, with 280 tons of nuts/year. Anzoátegui is the only state that has increased production, from 28 to 800 tons/year, with an average of 145 tons of nuts/year in this state. The state's Agricultural Research Center (CIAE Anzoátegui) and INIA-Agroforestry have carried out research projects with the maintenance of the clonal orchard of early dwarf cashew trees located at INIA Anzoátegui, with the aim of providing, in the short term, seeds and buds of materials of high yield and phytosanitary quality, destined for the establishment of up to 150 hectares of cashew plantations [15, 34].

In addition, INIA Anzoátegui has developed a selection of "creole" materials with promising characteristics to be used as rootstocks. Early dwarf clones from Brazil (CCP-76, CCP-1001 and CCP-06) have also been introduced, with excellent results in terms of the quality of the whole fruit (nuts and pseudo-fruit). In this way, the "criollos" have gradually been replaced by the early dwarf clones, increasing the area cultivated with this improved species by 534 hectares [15, 35].

The state of Amazonas has the highest yield, according to figures expressed in kg/plant. Finally, Zulia has 90 hectares and a yield of 160 kg/plant, where there are wild plantations, located mainly in rural settlements and some small plantations on farms as a secondary activity.

According to Figure 2.1, the state of Bolívar has the largest share, with 74.2% of production, followed by Guárico with 14.6%, Anzoátegui with 7.6%, then the state of Monagas with 2.7%, Amazonas with a minimal production of 0.7% and, lastly, the state of Zulia with a small production of 0.2%.

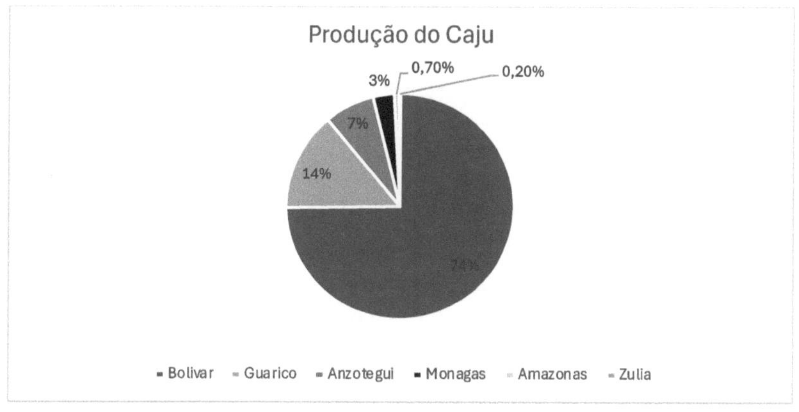

Figure 2.1 National cashew production in 2022

In order to find out about regional cashew production, data provided by the Corpozulia Fruit Center was taken into account. According to this Center, in the state of Zulia there is a planted area distributed between the municipality of Mara, with 50 productive hectares, and the municipality of Rosario de Perijá, with 40 developed hectares. Together, these municipalities have 90 productive hectares of cashew, which represents an annual production of 36 tons of nuts. The production of the municipality of Mara is shown below:

Table 2.5
Cashew acreage developed in the municipality of Mara

Parish	Cultivated area (ha)	Number of plants	Yield per plant	Nut yield (kg)
La Sierrita	7,49	749	8239,00	2059,75
Ricaurte	3,20	320	3.520,00	880,00
San Rafael	1,20	120	1.320,00	330,00
Tamare	14,65	1465	16.115,00	4028,75
Another municipality	6,13	613	6.743,00	1685,75
Total	32,67	3.267	35.937,00	8.984,00

Table 2.5 shows that Tamare parish has the largest cashew production in the municipality of Mara, with an area of 14.65 ha, corresponding to 1,465 plants, representing 16,115 kg of nuts and a total of 4,028.75 kg of kernels. Next, La Sierrita parish has an area of 7.49 ha, corresponding to 749 plants and a production of 82.39 kg of chestnuts and 20.6 kg of kernels. In a parish with no specific information, 6.13 ha are recorded, representing 613 plants, with a production of 6,743 kg of nuts and 1,686 kg of almonds. In Ricaurte parish, there is an area of 3.2 ha, corresponding to 320 plants and a production of 3,520 kg of chestnuts

and 880 kg of almonds. Finally, there is little cashew production in San Rafael parish, with 1.2 hectares and 120 plants, resulting in 1,320 kg of nuts and 330 kg of kernels, giving a total production in the municipality of 27,780 kg of nuts.

As for cashew production in the municipality of Rosario de Perijá, the so-called "Sabana Strip" stands out, which runs from Jalisco to the area known as 104. The sectors where cultivation is concentrated include: sector 2 de Fevereiro, San Juan I, San Juan, Guadalajara, La Patilla, El Carmen and La Matica I.

2.3.3 Imports

Imports are one of the variables to be studied in order to obtain apparent consumption, using the information provided by the National Statistics Institute (I.N.E.), Foreign Trade Statistics Consultation System. From this information, the statistics corresponding to imports were extracted.

Table 2.6 shows the behavior of cashew nut imports from 2015 to 2022. In the years 2015 to 2017, the trend is downward; then there is growth in the following year (2019), reaching significant growth in 2021, but falling again in the following three years: 2022, 2023 and finally 2024. In this sense, there is a cyclical behavior of the variable analyzed [32, 36, 37, 38, 39].

Table 2.6
National cashew imports

Country	kg	$	Bs	Bs/kg
Brazil	227.168,00	71.661,00	3.081.423,00	13,56
China	40.876,00	12.894,00	554.442,00	13,56
USA	2.672,00	842,00	36.206,00	13,54
Total	267.257,00	85.397,00	3.173.071,00	40,66

Figure 2.2 shows that the largest volume of cashew imports comes from Brazil (85%), which shows that this country has penetrated the Venezuelan market by selling dried cashew nuts. In second place, the USA, although it does not grow this product, has a 14% share of the Venezuelan market. The rest of the imports come from China, with a small share (2%).

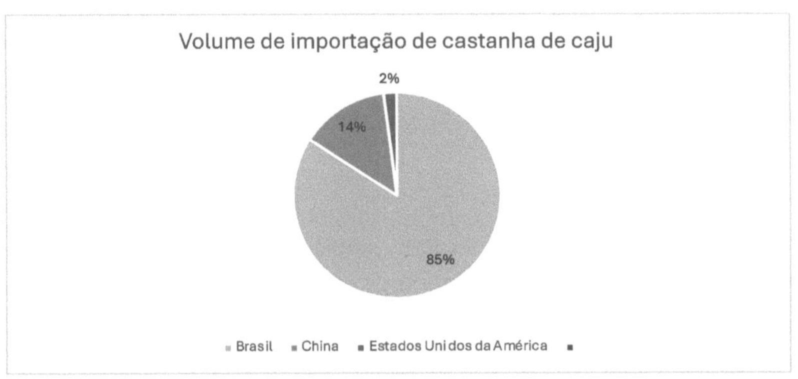

Figure 2.2 Cashew nut imports

2.3.4 Exports

Venezuela's exports were not very significant on international markets. In the period analyzed, only 370 kilos of cashews were exported to the Netherlands Antilles in 2002 and 11 kilos to Costa Rica in 2004. In this sense, exports are insignificant and their behavior has been decreasing to the point that, in the period from 2015 to 2024, there were no exports of the product analyzed [40, 41, 42].

Figure 2.3 shows Venezuela's international exports for the period 2015 to 2024, provided by the Foreign Trade Statistics Consultation System, where it can be seen that in 2017, 97% of cashew nuts were exported to the Netherlands Antilles and 3% to Costa Rica.

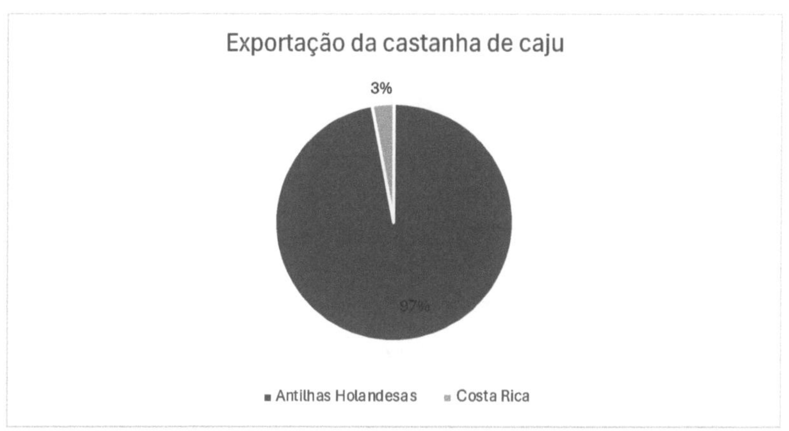

Figure 2.3 International cashew nut exports from Venezuela

2.3.5 Apparent national cashew consumption

Once all the variables that make up the so-called National Apparent Consumption (NAC) have been obtained, we proceed to calculate it, which represents historical demand. As mentioned above, National Apparent Consumption is defined as follows:

C.A.N. = P.N. (National Production) + I (Imports) - E (Exports)

This consumption allows us to know the history of cashew consumption at a national level.

Table 2.7 shows the National Apparent Consumption for each year analyzed. There is an increase when comparing the first year analyzed, 2015, with the last, 2024, from 1,859,375 kg to 2,229,657 kg, which results in a growth rate of 2.04% during the period analyzed, from 2015 to 2024.

Table 2.7
Apparent national cashew consumption

Years	Production (kg)	Import (kg)	Export (kg)	Apparent consumption (kg)
2015	2.044.200	185.457	0	2.229.657
2016	1.655.700	204.045	370	2.283.169
2017	1.770.450	137.500	0	2.337.965
2018	1.728.950	94.340	11	2.451.534
2019	1.921.950	158.860	0	2.510.371
2020	1.921.950	204.758	0	2.570.620
2021	1.921.950	565.487	0	2.632.315
2022	1.921.950	262.533	0	2.695.490
2023	1.921.950	267.282	0	2.760.182
2024	1.921.950	267.282	0	2.826.427

2.3.6 Estimated demand

The estimated national cashew demand for the period 2015 to 2024 was calculated based on the projection of National Apparent Consumption for the historical period 2015 to 2024, as shown in Table 2.7, applying the growth rate which, in this case, resulted in r = 2.04%.

Figure 2.4 shows how projected demand increases each year, in line with the growth rate (r = 2.04%) observed in apparent consumption.

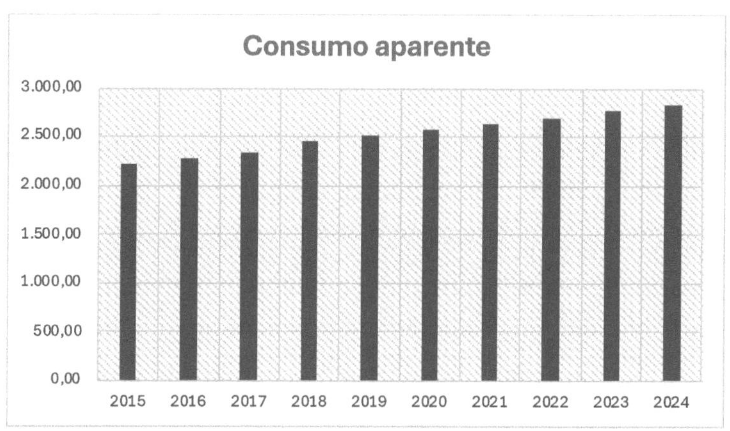

Figure 2.4 Estimated national demand for cashew nuts

2.3.7 Supply analysis

2.3.7.1 World cashew nut supply

In the early 1970s, most of the global cashew production (68% of the total) took place in African countries, particularly Mozambique and Tanzania. Later, in 1999, this trend changed, with Asian countries emerging as leaders, reaching 57% of world cashew production. That year, production was as follows: India with 430,000 t, Brazil with 140,323 t and Vietnam with 70,000 t.

In 2001, India continued to lead production, but with a smaller share, around 40% of the world market. Over the next seven years, Asian countries continued to dominate production, but now other Asian countries, especially Vietnam, and Nigeria have begun to expand their production capacity. In 2022, Table 2.8 shows the world production of cashew nuts [32, 33, 43].

Table 2.8
World cashew supply (merey) (2022)

Country	Production (t)
Ivory Coast	970.000
India	752.000
Vietnam	341.680
Benin	215.000
Indonesia	163.083
Brazil	147.137
Mozambique	144.823
Guinea-Bissau	116.484
Nigeria	74.690
Tanzania	24.922

2.3.7.2 National cashew nut supply

Supply is analyzed to measure production volumes and the conditions for making cashew production available on the market. To offer any product, several factors must be taken into account, such as prices, private or government financial support and competition. For the projected supply, consideration must be given to the existence of expansion of raw material production or new crops, as well as plans to expand new and existing plants.

In this case, there is no knowledge of the expansion of new wild-type crops, so national production from 2022 to 2030 will be taken into account to forecast supply.

To do this, national production is projected and the growth rate method is used, which has been set at 2.37% (Table 2.9).

Table 2.9
Projected supply of national production

Year	Offer (kg)

2022	2.092.641
2023	2.142.229
2024	2.192.993
2025	2.244.960
2026	2,298.158
2027	2.352.617
2028	2.408.366
2029	2.465.436
2030	2.523.859

2.3.8 Available national demand

Available national demand is obtained from the difference between projected demand and projected supply. Table 2.10 shows the available national demand.

Table 2.10 shows that the demand for cashew nuts is greater than the supply, which indicates that there is an available demand, and therefore an unsatisfied market.

Table 2.10

Available national demand

Year	Projected demand (kg)	Offer (kg)	Available demand (kg)
2022	2.275.105	2.092.641	182.465
2023	2.321.480	2.142.229	179.251
2024	2.368.800	2.192.993	175.807
2025	2.417.085	2.244.960	172.125
2026	2.466.353	2.298.158	168.195

2027	2.516.627	2.352.617	164.010
2028	2.567.924	2.408.366	159.558
2029	2.620.268	2.465.436	154.831
2030	2.673.678	2.523.859	149.819

2.4 Price Analysis

The prices considered are the same as those currently established on the market. According to interviews with traders, distributors and wholesalers, the price is Bs. 114 per kilo.

2.5 Marketing

The investigations carried out determined that all production at national level is destined for domestic consumption and the cottage industry. In fact, marketing is restricted to the localities where the fruit is produced (Villa del Rosario, Mara) and the municipality of Maracaibo.

It is sold in different presentations: cashew in syrup, a typical marzipan in the east of the country and the famous roasted cashew, which is the most popular presentation due to its practicality, durability and easy handling of the finished product, which is also the presentation analyzed in this project [44, 45].

Despite improvements in cultural practices for the development and productive efficiency of cashew, in our country the main objective of exploitation continues to be the kernel, while the pseudofruit remains underutilized. This results in losses at the marketing stage, since the pseudofruit or pulp, which represents around 90% of the edible portion of the cashew, has great potential for fresh or processed consumption. This project envisages the use of aluminized packaging with a capacity of 500 grams. Cashew marketing is very irregular and is practically restricted to the production areas, however, it is hoped to market at a site that will sell, in addition to the kernels, cashew derivatives and Wayú handicrafts.

CHAPTER III: SIZE AND LOCATION

3.1 Size

3.1.1 Factors that determine size

To determine the size of the plant, criteria related to the description and production of the raw materials were analyzed, as well as the effective capacity of the equipment [46, 47, 48, 49].

3.1.1.1 Description of the raw material

The fruit of the cashew is the nut, made up of shell, oil and kernel. The kernel, which is the actual seed, represents 25% of the nut, has a curved shape, slightly kidney-shaped, white when natural, and golden when fried, weighing 1 to 2 grams.

To feed the cashew processing plant, 90 ha are being considered, distributed as follows: 50 ha in the municipality of Mara and 40 ha in the municipality of Villa del Rosario.

The yield of the total 90 ha, with a density of 100 plants/ha, is 9,000 plants. The yield per plant is 160 kg, giving a total of 1,440,000 kg of fruit. Considering that the trees are over 20 years old and that the nut represents 25% of the weight of the fruit, the total nut production is 360,000 kg.

3.1.1.2 Cashew nut production program

The chestnut production program was estimated considering the number of hectares planted and taking into account a yield of 16 t/ha, which results in a total production of 1,440,000 kg of fruit per year. Considering that the nut represents 25% of the weight of the pseudofruit, in the first year, production is 360,000 kg of nuts. Then the kernel, which represents 25% of the nut, results in 90,000 kg of kernels per year, remaining constant throughout the period analyzed [35, 44].

3.1.1.3 Effective Equipment Capacity

The plant's installed capacity is 500 kg of nuts per day. Processing 500 kg of cashew nuts translates into 125 kg of almonds per day.

For the first year, 80% of the installed capacity will be processed per day, which translates into 100 kg of nuts per day, representing 28,800 kg of nuts per year, with an increase year after year, until it stabilizes in year 3, when 100% of the capacity will be used, processing 500 kg of nuts per day, from which 36,000 kg of nuts per year will be obtained.

3.1.1.4 Cashew kernel production program

The almond production program depends on the hectares needed to supply the processing plant. For a year's work, production from 36 ha is required, which translates into 144,000 kg of nuts per year, in order to achieve 36,000 kg of almonds per year.

3.2 Location

The pilot cashew processing plant will be located on land that is part of the "Comprehensive Producer Care Center", in the Los Haticos sector, next to the El Carmen neighborhood, entering from Altos de Jalisco, in the municipality of Rosario de Perijá.

The area earmarked for this pilot plant will be located in the same region as the plantation, for various reasons, the main one being the timely transportation of the raw material to the processing site. In addition, the lack of job opportunities in the municipality forces its residents to migrate to other municipalities.

3.2.1 Factors influencing location

The location of a processing plant is mainly influenced by its proximity to cultivated areas. In this case, the plant will be supplied with raw material from the Mara region, coming from plantations in the following parishes: Tamare, with 14.65 ha, La Sierrita, with 7.49 ha, Ricaurte parish, with an area of 3.2 ha, and San Rafael parish, as well as the area of the municipality of Rosario de Perijá, specifically in the "Franja de Sabana", from Jalisco to the area known as 104. The sectors where the crop is concentrated include: sector 2 de Fevereiro, San Juan I, San Juan, Guadalajara, La Patilla, El Carmen and La Matica I (the area close to where the pilot plant will be installed).

Another important factor is the access routes and availability of public services (water, electricity and gas) in the region, availability of labor (people from the community with experience in the area will be incorporated to act as plant operators), proximity to supply sources, environmental factors, proximity to the market, availability of land and topography of the soil.

CHAPTER IV: TECHNICAL ASPECTS

4.1 Description of the production process

Cashew nut processing can be summarized in four main stages:

4.1.1 Receiving and weighing, drying the chestnuts, cleaning and washing, sorting and cooking the chestnuts (process and autoclave, cooling and drying the chestnuts)

Receiving and weighing:

When the raw material (nuts) arrives in sacks, it is weighed to settle the payment. It is then stored in piles and dried.

Drying the nuts:

Drying should be done on flat or cemented platforms. The chestnuts should be exposed to the open air, in the sun, for around 72 hours (3 days), reducing the humidity to between 7% and 9% to avoid problems of deterioration, especially by fungi. This process takes place in cement blocks or on the patio, depending on the region. In humid regions, the chestnuts should be covered at night with a tarpaulin or plastic wrap to prevent moisture from dew, or kept under cover to protect against rain. The chestnuts should be spread on cemented soil with a maximum thickness of 7 to 10 cm, allowing for aeration and sunlight. It is recommended to plow the soil and turn the kernels over twice a day. Cashew nuts are harvested between August and December, so processors need inventory to work with all year round.

Cleaning and washing:

The nuts are then taken to the cleaning area, where stones, sand, remains of pseudo-fruit and other impurities are removed by hand. The cleaning process follows the criterion of removing all contaminating impurities (earth, stones, leaves, insects, etc.) to avoid contamination and deterioration. The cleaned nuts are bagged and placed on wooden pallets. It is important to avoid stained, damaged, deteriorated, malformed or rotten kernels, which could contaminate the batch.

Classification:

This process is carried out by a rotary sizer and sorter. At this stage, the nuts are sorted by size. Sorting is done with sieves, and the nuts pass through perforated cylinders of different sizes, according to the sorting criteria (small, small, medium, large and extra), appropriate to the market. The proper classification of the nuts or almonds is essential for the success of the next operations, such as the penetration and uniform absorption of heat in the autoclave and cutting in the manual cutting machines, facilitating the work of shelling and the process of frying or roasting the almonds. This task can also be done manually in artisanal productions.

Process and autoclave:

The nuts are then taken to the cooking area, where they remain for 15 minutes in an autoclave. The autoclave is a steam generator that expands the shell in relation to the inner kernel, hardening the shell while the kernel remains the same size, making it easier to cut without breaking it and preserving its shell. This process minimizes the spillage of LCN (liquid from the nut shell), avoiding contact and contamination of the kernel. They are then taken to the cooling area and placed on trays.

Cooling and drying the nuts:

After the autoclave process, the nuts should rest in a suitable place until they reach room temperature. For greater productivity, it is recommended that they be placed in greenhouses or dehydrator ovens for faster and more uniform drying, providing a satisfactory cut for the manual peelers. They are then transported manually in buckets to the peeling area.

4.1.2 Cutting the Nut Shell, Dehydrating and Swelling the Kernel

Cut off the shell of the chestnut:

At this stage, the shell is cut off, obtaining the nut shell as a residue and the wet kernel as the main product (which will be processed until the roasted kernel is obtained). The wet kernels are transported in buckets to the drying area.

Almond dehydration:

The wet kernels are dried in an oven at a temperature of 70 to 80 °C. After the oven, the roasted kernels are taken to the depulping area and cooled.

Almond cracking:

This process can be done by hand, on tables with appropriate trays, until it reaches room temperature. The moisture content of the kernels is reduced to between 2.5% and 3%, making the film previously adhered to the kernel fragile and making it easier to remove. Drying takes place in ovens with a hot air flow (60 °C to 70 °C) for 6 to 8 hours. The kernels are placed on trays and heated so that the film is loosened evenly. In this project, the kernels undergo a humidification process with saturated steam (1-2 minutes) to facilitate the separation of the kernel from the skin.

4.1.3 Kernel shelling, sorting and grading

Peeling the kernels:

Depulping can be done in various ways. At this stage, operators use a simple finger movement to separate the film from the kernel. In some cases, by rolling a rope over a metal blade, parts of the adhered film can be removed.

This project will use a dehuller with a manual tray and a bristle brush, with a production capacity of 15 kg per operator. This equipment must be handled with great care to avoid a considerable increase in the percentage of broken kernels.

Selection and classification:

At this stage, the kernels go to the sorting area. If the kernels are to be sold naturally, after sorting, the packaging process follows. The kernels are classified mainly by size, integrity and color. The operation is carried out on tables with benches covered in formica or thick, light-colored cloth, where the kernels are treated on a smooth surface that retains the dust. The aim is to select 75% whole kernels and 25% broken kernels for storage and marketing.

4.1.4 Toasting and packaging

Toasting:

To sell the fried almonds, the almonds must be fried separately by size to ensure even frying. The equipment used can be the same as that used to fry potatoes. In this case, a gas roaster will be used.

It is advisable to use good quality oil to avoid unwanted flavors. The most commonly used oils are those derived from corn or soy.

The recommended procedure for frying and curing is as follows:

- The kernels of similar size and color are placed in appropriate baskets and dipped in hot oil at the ideal temperature for frying. The amount of oil should be enough to cover all the kernels and avoid uneven frying.

- The frying time varies from 3 to 6 minutes, depending on the volume of almonds in the baskets. It is recommended not to stir the almonds to prevent them from breaking.

- After frying, excess fat is removed from the product by pouring the contents of the basket onto a flat surface covered with absorbent paper or a clean jute bag. The best result is obtained using a centrifuge.

- Curing is carried out while the kernels are still hot, using good quality refined salt, dry and free of impurities, in an amount of 1% to 2% of the kernel's weight. The kernels should be separated by size to allow uniform frying.

Packaging:

In this project, the finished product will be marketed at the retail level, with the almonds being packed in low-density plastic bags or metal bags with hermetic aluminum seals. The product will be packaged in units with a net weight of 250 to 500 grams. The container used for the almonds must be new, clean, dry, water-resistant, lead-free and hermetically sealed. The container must also be strong enough to guarantee the integrity of the product during transportation and storage.

4.2 Equipment description

Machinery and equipment:

The machinery and equipment were selected on the basis of the production process, the size of the project and the technology available. The main pieces of equipment are described below:

Platform scale: Scale from 50/100 g to 150/200 kg for weighing the quantity of nuts received and measuring the quantity of almonds to be dispatched per package.

Gas Autoclave: Manual, with a capacity of 5000 liters, equipped with a processing tray and loading mouth. Its dimensions are 0.6 x 0.60 x 1.30m. It is a steam generator that promotes the growth of the nut in relation to the kernel, as well as the hardening of the shell. While the nut expands, the kernel maintains its size, making it easier to cut without breaking and preserving the skin. The autoclave minimizes liquid spillage from the shell, preventing contamination of the kernel (50/80/100 kg every 25 minutes).

Airing rack and trays: Made of stainless steel, used to cool the almonds to room temperature. The almonds are spread out on the trays. The rack should have wheels to facilitate transportation within the plant.

Shelling machine: Manual, with a capacity of 45 kg/operator, equipped with a screen, universal cutting handle and pedal (dimensions: 0.8 x 0.30 x 0.30 m). Adjusted to the types of nuts classified, it is fixed to stainless steel workbenches. The operation must be carried out by two operators: one cuts the nut and the other removes the kernel from the shell.

Double Cutting Table: Made of carbon steel, equipped with 3 compartments for installing the nut peeler, with a capacity of 50 kg (dimensions: 2.2 x 0.55 x 0.60 m).

Double Dehydrator for Almonds: With electric gas burner, equipped with 20 trays and 2 fans with 1/3 hp electric motors (dimensions: 1.60 x 0.75 x 1.70 m). The LPG-powered oven reduces the almonds' humidity, making it easier to remove the skin. The hot air (70ºC) circulates through the removable trays for 5 to 7 hours (30/50/70/100 kg every 6 hours).

Humidifier: Equipped with an evaporation chamber with 5 fabric trays (dimensions: 0.60 x 0.75 x 1.30 m). The steam/saturation chamber subjects the kernels to a saturated steam humidification process for 4 to 7 minutes to facilitate the removal of the kernel skin (21 kg of seeds every 5 minutes).

Dehuller: Manual, equipped with a tray and brush for removing the film, with a production capacity of 15 kg/operator (dimensions: 0.80 x 0.60 x 0.60 m).

Frying equipment: Consists of 1 LPG fryer and a centrifuge for drying the kernels, with a reinforced structure (dimensions: 0.60 x 0.55 x 0.40 m). The salt is applied while the kernels are still hot.

Sorter: Made up of a rotating sieve, in which the nuts are sorted by size, passing through perforated cylinders of different sizes, according to the sorting criteria.

Packaging equipment: Comprising a bench, fed by a tolva and a sealer for packaging the final product (Dimensions: 0.40 x 0.55 x 0.40 m). The fried and salted product is packed in low-density plastic bags or metal bags with hermetic aluminum seals. The net weight of the package varies from 100 to 400 grams.

4.3 Cashew nut process line

In chemical engineering, an industrial process line refers to a sequence of interconnected unit operations that transform raw materials into end products. This sequence can include a series of steps, each with its own function and purpose.

Each of these stages can have specialized equipment, and the design of the process line must consider optimizing the efficiency, safety and sustainability of the process.

The process line for making the final cashew kernel product is shown in Figure 4.1 and Figure 4.2.

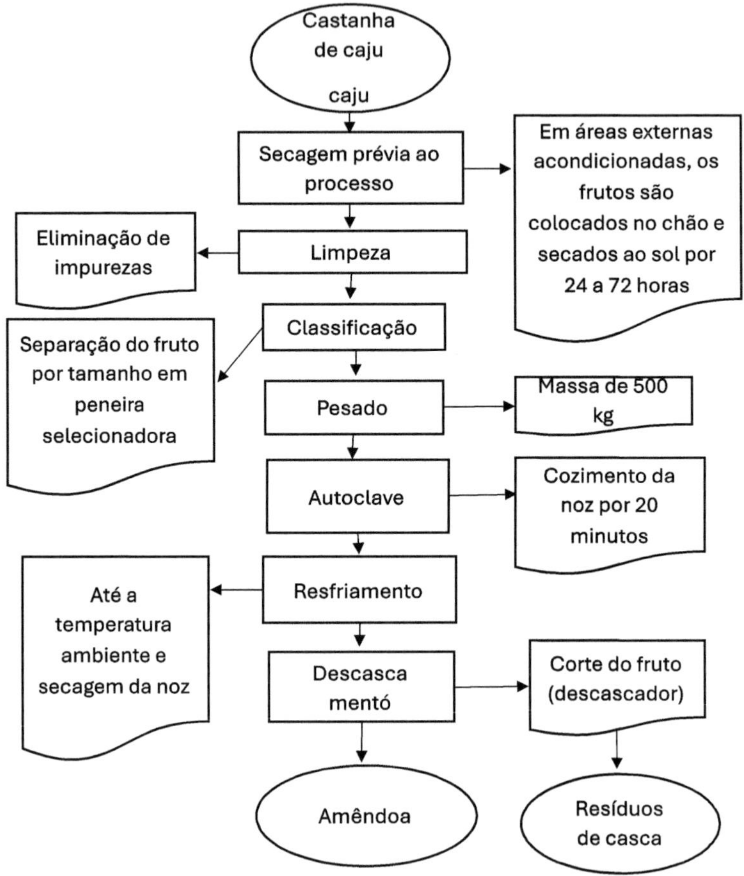

Figure 4.1 Flow diagram of the production process from cashew nut to cashew kernel

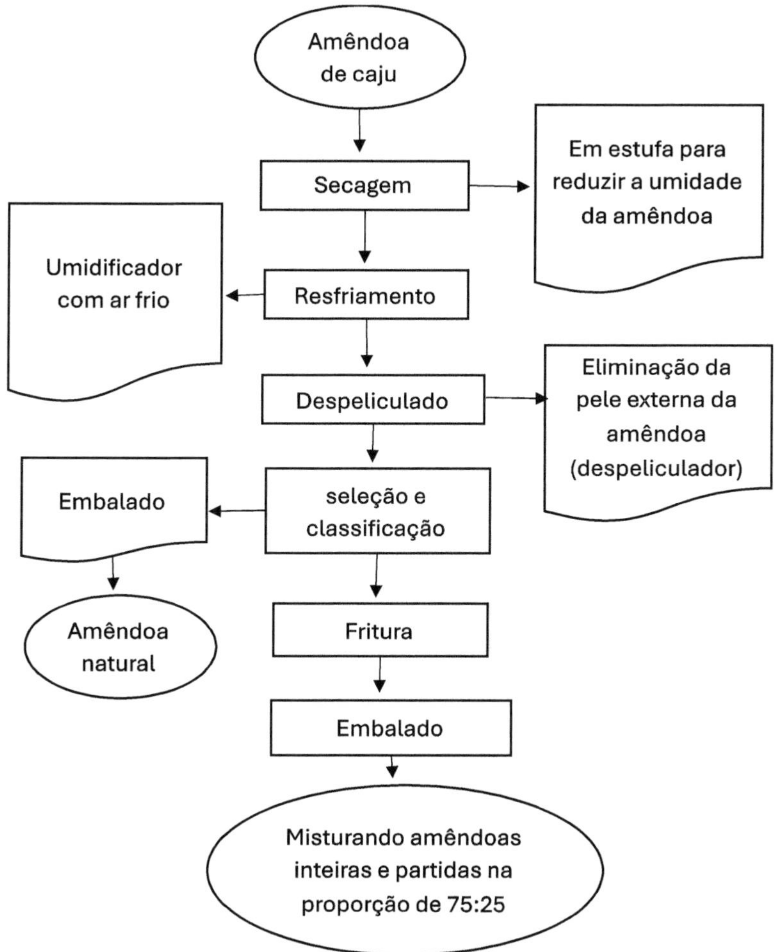

Figure 4.2 Flow diagram of the production process from the kernel to the final cashew product

4.4 Balance of Materials

Material balance is a key process in industrial, chemical and process engineering. It consists of calculating and analyzing the quantities of materials entering and leaving a system, ensuring that the law of conservation of matter is complied with. This balance is fundamental for optimizing processes, managing resources and ensuring efficiency in operations.

According to the structure of the nut and the weight ratio of its components (shell 49%, oil 23%, moisture and impurities 3% and kernel 25%), during the processing of 500 kg of nuts to obtain 36,000 kg/year of kernels at maximum plant capacity, around 245 kg of shell are generated, 115 l of caustic oil from the shell of the real fruit, 15 l of water and 125 kg of kernel, as shown in Figure 4.3 below:

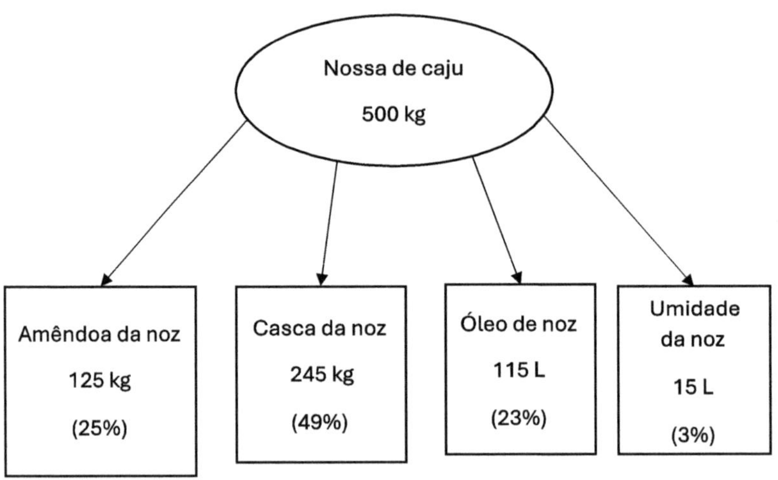

Figure 4.3 Material balance of the walnut structure and the weight ratio of its components

4.5 Distribution of areas by process

The operations for obtaining almonds can be carried out in six different areas as follows:

- First, you need to set aside a space for receiving and weighing the nuts.

- Secondly, the production area: A space of approximately 0.60 x 0.60 m should be allocated in the production area, where the autoclave will be located. At this stage, the nuts are boiled and cooled and then the shells are removed.

- A space should be set aside to position the nut-cutting machine, with its 2.2 x 0.55 m table.

- Space for storing the shells, since at this stage the almonds are cut and the waste generated is the shells.

- In the production area, another space should be set aside to place the 1.60 x 0.75 m steam oven. This stage reduces the almonds' humidity to 68%, making it easier to remove the skins.
- In the production area, space must be allocated to place the humidifier, which requires a space of 0.60 x 0.75m. This step also aims to reduce the humidity of the almonds, making it easier to remove the skins.

- In the production area, space must be allocated to position the depelicator, which requires a 0.80 x 0.60 m space. This stage removes the films.

- Space is needed to position the almond depulping table close to the nut cutting area.

- As well as a table for sorting the almonds.

4.6 Raw material requirements

In the first year, 80% of the installed capacity will be processed, so 400 kg of nuts will be needed per day. In the second year, 450 kg of nuts will be needed per day, until stabilizing in the third year, when the plant will operate at 100% of installed capacity, processing 500 kg of nuts per day. Considering that 288 days per year will be worked, a total of 144,000 kg of nuts will be needed per year.

4.7 Packaging material

The product will be presented for sale in packaging consisting of aluminized bags with a capacity of 500 grams. For the first (1st) year, a total of 57,600 units will be needed; for the second (2nd) year, 64,800 units, until reaching 72,000 units from the third (3rd) year onwards, remaining constant until the last year analyzed.

4.8 Manpower requirements

For the normal operation of the cashew mini-processing plant, the following personnel will be needed: four (4) operators in the receiving area, six (6) operators in the cutting area, as six (6) double cutting tables will be needed, five (5) operators in the depelletizing area, four (4) operators responsible for sorting and selection, three (3) operators in the packaging area, three (3) additional workers distributed among the different areas that will be set up for processing, three (3) humidification operators.

In addition, for the smooth running of the plant in the administrative area (offices), it will be necessary to have one (1) plant coordinator, one (1) quality supervisor, two (2) cleaning workers, one (1) security guard, two (2) administrative assistants, one (1) accountant and one (1) secretary.

4.9 Office furniture and equipment

Office furniture and equipment consists of all the equipment and tools needed for the plant to function properly.

4.10 Quality control

Quality assurance involves planning and monitoring quality in a company or organization. The main objective of quality assurance is to generate trust both internally and externally.

Quality control is fundamental to the successful production of excellent walnuts, and it is necessary to apply quality controls at various stages of the process, especially when receiving the raw material and during the key cooking and cutting operations.

Receiving raw materials:
Receiving the raw material must involve checking the characteristics of the nuts as an indispensable condition for obtaining a good quality kernel. At this stage, the appearance of the nut is inspected in terms of weight, size, shape, color, texture, as well as the presence of foreign material that could affect the yield of the process.

Cooking and cutting operations:
During cooking and cutting operations, the following controls are required:

Cooking:
Strict temperature and time control must be maintained to ensure uniform heating of the nuts, which is essential for proper separation of the shell and kernel in the cutting operation.

Cutting:
It's important to adjust the intensity of the cut according to the size of the nut, so that too much pressure doesn't cause the kernel to break or too little pressure makes it difficult to separate the shell from the kernel.

Final stage of the process:
At the final stage of the process, it is essential to obtain a kernel that meets the quality attributes required for this type of product. Once the kernels

have been obtained, it is necessary to assess the quality of the final product, which will be determined by the condition of the fruit when it arrived at the plant and the processes applied. If the processes have been carried out properly, maintaining hygiene in each operation, the resulting almond will have acceptable and satisfactory quality characteristics. In this sense, the following parameters should be checked:

Characteristic smell, whole kernels, light brown, dry, free from bites, defects or blemishes.

- Humidity = 10%
- Crude protein (N 6.25)* = 29.9%
- Saturated fatty acids = 18.5%
- Unsaturated fatty acids = 81.5%
- Total carbohydrates = 27.2%
- Crude fiber = 1.2%
- Mineral salts = 1.7%
- Calcium = 165 mg/100g
- Phosphorus = 490 mg/100g
- Iron = 5 mg/100g
- Thiamine = 140 mg/100g
- Riboflavin = 150 mg/100g
- Total nicotinic acid = 2.2 mg/100g

4.11 Basic Services

The basic services that the cashew mini-processing plant must have in order to function properly are: water, electricity, gas and telephone.

4.12 Maintenance

Plant equipment will be maintained in accordance with the manufacturer's specifications. However, it is necessary to carry out periodic checks on the oil levels of the installed equipment in order to prevent failures. In addition, periodic checks must be made on the electrical system to prevent system failures.

CONCLUSIONS AND SUGGESTIONS FOR FUTURE WORK

When laying out the pilot plant, a number of considerations need to be taken into account to allow for an efficient arrangement of equipment and space in general. Furthermore, this comprehensive approach will not only improve cashew nut production, but also provide valuable information on the economic viability and sustainability of the process .

- Making the most of the available space, minimizing travel distances and optimally distributing corridors, warehouses, equipment and employees.

- Trying to ensure that the material remains in the process for as short a time as possible, reducing distances in the distribution of equipment and creating logical production sequences, preferably in a linear direction.

- Supervision of the entire process is extremely important at all times, which is why the field of vision for supervision should be planned in the design so that as many of the processes involved as possible can be seen in a single view.

- Avoid congestion in the aisles, as well as machines standing still waiting for material to be processed. This can be avoided with a linear or L-shaped distribution of production equipment, making good use of the operating times of each piece of equipment, or between one stage and the next. Likewise, by considering this measure, the rates of accidents, incidents or sudden changes in trend are minimized.

- The following distribution of equipment in production should be designed and, in general, a suggested distribution for the entire plant, in order to maximize production efficiency while minimizing indirect costs due to lost time, both due to idle equipment and idle personnel.

- To contribute to the development of the cashew nut agro-industrial industry by introducing more efficient and sustainable processes.

- Provide valuable information for producers and companies interested in large-scale production.

BIBLIOGRAPHICAL REFERENCES

[1] Ferrer, J., & Alejos, R. (2024). Science and technology of cashew agro-industrial processing. London, United Kingdom: New Academic Editions.

[2] Central American technical regulation 67.01.33:06. Processed food and beverage industry. Good Manufacturing Practices. General Principles. Adaptation of CAC/RCP-1-1969. Revision 4-2003. Recommended International Code of General Food Hygiene Practices. Available at https://asp.salud.gob.sv/regulacion/pdf/rtca/rtca_67_01_3306_bebidas_procesadas_buenas_practicas.pdf

[3] Codex Alimentarius international food standards, guidelines and codes of practice contribute to the safety, quality and fairness of this international food trade. Available at https://www.fao.org/fao-who-codexalimentarius/en/

[4] Code of hygienic practice for tree nuts (CAC/RCP 6-1972). Codex Alimentarius international food standards.

[5] Cardozo, R. (2022). Modeling the kinetics of osmotic dehydration and kiln drying of the Acadardium occidentales L. pseudofruit for raisin production in Zulia state. (Master's Thesis in Food Science and Technology). University of Zulia. Venezuela.

[6] Garruti, D. de S., Lima, J. R., Lima, A. C., Paiva, F. F. de A., Barros, M. E. S., & de Moraes Í. V. M., Pinto de Abreu, F. A , Machado, T. F., Rocha Bastos, M. do S., da Silva Neto, R. M., de Souza Filho, M. de S. M., & Nassu, R. T. (2015). Industrial Utilization. In J. P. P., de Araújo (Ed.) Cashew (pp. 188-238).

[7] General principles of food hygiene (CXC 1-1969). Adopted in 1969. Amended in 1999. Revised in 1997, 2003, 2020. Editorial corrections in 2011. Codex Alimentarius international food standards.

[8] Food Hygiene (2006). Basic Texts / Pan American Health Organization; National Health Surveillance Agency; Food and Agriculture Organization of the United Nations. Brasília.

[9] Ferrer, J., & Silva, V. (2018). Sistema de análisis de peligros y puntos críticos de control. Mauritius. Editorial Académica Española.

[10] PAS field project. CNI/SENAI/SEBRAE/EMBRAP agreement. (2004). Manual of Good Agricultural Practices and HACCP System (Food Quality and Safety Series). Brasilia: EMBRAPA/SEDE.

[11] SQF Food Safety Code: Food Manufacturing (2020). (Edition 9) SQF Institute. FMI The Food Industry Association. Arlington, VA. USA.

[12] Guerrero, R., Lugo, L., Marín, M., Beltrán, O., De Pinto, G. L. & Rincón, F. (2008). Physicochemical characterization of the fruit and pseudofruit of Anacardium occidentale L. (merey) under drought conditions. Revista de la Facultad de Agronomía, 25(1), 81 - 94.

[13] Lemos, Moaciria de Souza, Bordallo, Patricia do Nascimento, Vidal Neto, Francisco das Chagas, Lima, Eveline Nogueira, & Holanda, Ioná Santos Araújo. (2021). Genetic and physicochemical diversity in sweet Brazilian cultivars of anacard (Anacardium occidentale). Acta botánica mexicana, (128), e1775. https://doi.org/10.21829/abm128.2021.1775

[14] Sivagurunathan, P., Sivasankari, S., & Muthukkaruppan, S. (2010). Characterization of cashew apple (Anacardium occidentale L.) fruits collected from Ariyalur District. Journal of biosciences research, 1(2), 101-107.

[15] Alejos, R. (2012). Proyecto para la instalación de una miniplanta procesadora de merey en el Municipio Rosario de Perija. CESID-Frutícola e Apícola. CORPOZULIA. Venezuela.

[16] Cashew Production System (2016). Production System Data. Embrapa Agroindústria Tropical. Production System, 1. Electronic Version. (2nd edition). 1-193. https://www.spo.cnptia.embrapa.br/conteudo?p_p_id=conteudoportle...

[17] Prommajak, T., Leksawasdi, N., & Rattanapanone, N. (2014). Biotechnological Valorization of Cashew Apple: A Review. Chiang Mai University Journal of Natural Sciences, 13(2), 159-182.

[18] Nomwendé, J., Toulsoumdé, L., Vianney, W., Bationo. F., & Hama, M. (2021). Morphological characterization and quality assessment of cashew (Anacardium occidentale L.) nuts from 53 accessions of Burkina Faso. Journal of Agriculture and Food Research, 6, 100219. https://doi.org/10.1016/j.jafr.2021.100219

[19] Pereira, A. L. F., Maciel, T. C., & Rodrigues, S. (2011). Probiotic beverage from cashew apple juice fermented with Lactobacillus casei. Food Research International, 44(5), 1276-1283. https://doi.org/10.1016/j.foodres.2010.11.035

[20] Cruz Reina, L. J., Durán-Aranguren, D. D., Forero-Rojas, L. F., Tarapuez-Viveros, L. F., Durán-Sequeda, D., Carazzone, C., & Sierra, R. (2022). Chemical composition and bioactive compounds of cashew (Anacardium occidentale) apple juice and bagasse from Colombian varieties. Heliyon, 8(5), e09528.

[21] Cashew coast (2024). The role of food safety, quality and fair standards in cashew nut processing. https://139839173.hs-sites-eu1.com/welcome-to-the-cashew-coast-1

[22] Stefano, B. D. (2014). Cashew, from seed to market: a review. Agronomy for Sustainable Development, 34(4), 753-772. DOI: https://doi.org/10.1007/s13593-014-0240-7

[23] Chunilal, G. J. (2022). Cashew Nut Processing Unit. Arts, Science, Commerce College, Uttam Nagar CIDCO, Nashik 08 (Affiliated to Savitribai Phule Pune University, Pune).

[24] Oliveira, N. N., Mothé, C. G., Mothé, M. G., & de Oliveira, L. G. (2020). Cashew nut and cashew apple: a scientific and technological monitoring worldwide review. Journal of Food Science and Technology, 57(1), 12-21. doi: https://doi.org/10.1007/s13197-019-04051-7

[25] Al Mani, S , & Yudha, E. P. (2021). The Competitiveness of Indonesian Cashew Nuts in The Global Market. JEJAK Journal of Economics and Policy. 14 (1), 93-101. doi: https://doi.org/10.15294/jejak.v14i1.26067

[26] Azam-Ali, S. H., & Judge, E. C. (2001). Small-scale cashew nut processing. FAO. ITDG Schumacher Centre for Technology and Development Bourton on Dunsmore, Rugby, Warwickshire, UK. http://www.fao.org/ag/ags/agsi/Cashew/Cashew.htm

[27] Fitzpatrick, J. (2011). Cashew Nut Processing Equipment Study - Summary African Cashew initiative. Deutsche Gesellschaft für Internationale Zusammenarbeit (GIZ) GmbH International Foundations Postfach 5180, 65726 Eschborn, Germany.

[28] Guidebook on the Cashew Processing Process (2024). Crunchy Cashew. https://crunchycashews.com/about-us

[29] Vietlinh (2023). Standards for grading cashew nuts for export. Vietlinhagrimex CO., LTD.

[30] African Cashew Initiative (GIZ). (2020). Knowledge platform on family farming (How to estimate the quality of fresh cashew nuts (RCN)). Food and Agriculture Organization of the United Nations FAO TECA. Ghana. Africa. https://www.fao.org/family-farming/detail/en/c/1619104/

[31] Mordor Intelligence (2024). Anacard market share and size analysis: trends and growth forecasts (2024-2029). https://www.mordorintelligence.com/es/industry-reports/global-cashew-market

[32] Fresh/Dried Cashew Nuts, Product Trade, Exporters and Importers. The Observatory of Economic Complexity. (2022). https://oec.world/en/profile/hs/freshdried-cashew-nuts

[33] Market price of cashew nuts in the United States: prices and graphs (2024). Freshela exporters. https://www.freshelaexporters.com/cashew-nuts/prices/the-united-states/

[34] Plan for the Massification of Scientific Routes of the National Scientific Seeds Program (2024). Directorate for the Application and Generation of Knowledge. Pilot plant of the State Research Center for Experimental Agroindustrial Production (CIEPE). Venezuela.

[35] Cashew nut processing (2024). Practical Action. Technology challenging poverty.The Schumacher Center for Technology & Development. UK.

[36] Rossetti, A. G., Vidal Neto, F. das C., & Barros, L. de M. (2019). Sampling of cashew nuts as an aid to research for the genetic improvement of cashew tree. Pesquisa Agropecuária Brasileira, 54, e00962. DOI: https://doi.org/10.1590/S1678-3921.pab2019.v54.00962.

[37] Antônio Calixto Lima, Francisco das Chagas, Vidal Neto Carlos Wagner, Castelar Pinheiro Maia, Pedro Felizardo, Adeodato de Paula, Pessoa Francisco & Fábio de Assis Paiv. (2022). Recommendations for the Rapid Evaluation of the Quality of Industrial Cashew Nuts. Fortaleza, CE. Brazil.

[38] Castro, A.C.R.d.; Barbosa, E.R.; Cruz, A.C.A.d.; Segundo, V.C.V.;Pereira, M.A.; Lima, A.C.; Torres, C.R.B.; Aragão, F.A.S.d.

Characterization of Cashew Nut (Anacardium occidentale L.) Germplasm for Kernel QualityAttributes. Int. J. Plant Biol. 2023, 14, 1092-1099. https://doi.org/10.3390/ijpb14040079

[39] Cevallos, M., Urdaneta, F. and Jaimes, E. (2019). Development of agroecological production systems: Dimensions and indicators for their study. *Revista de Ciencias Sociales (Universidad del Zulia, Venezuela), XXV* (3), 172-185.

[40] Alejos-Pineda, R. E. (2003). Physical-chemical characterization of the fruit of two types of merey (*Anacardium occidentale L.*) for industrialization purposes. (Master's Thesis in Food Science and Technology). University of Zulia. Venezuela.

[41] Alejos-Pineda, R., Arenas de Moreno, L., Ferrer, J., Castellano, G., Nuñez-Castellano, K. and Pérez-Pérez, E. (2022). Physical-chemical characterization of the fruit of two types of merey (*Anacardium occidentale L.*) from a plantation in Mara, Zulia state, Venezuela. Revista Iberoamericana de Tecnología Postcosecha, 23 (2), 166-180. Available at: https://www.redalyc.org/articulo.oa?id=81373798007

[42] Sindoni, M., Loggiodice, P. R. H., & Natera, J. R. M. (2009). El merey (Anacardium occidentale L.): La especie frutal de las sábanas Orientales de Venezuela. Revista Científica UDO Agrícola, 9(1), 1-8. Available at: https://dialnet.unirioja.es/servlet/articulo?codigo=3293540

[43] McLaughlin, J., Balerdi, C. and Crane, J. (2022). The marañón (*Anacardium occidentale L.*) in Florida. Horticultural Sciences, Florida Cooperative Extension Service, Institute of Food and Agricultural Sciences, University of Florida (UF/IFAS), HS1041, 1-4. https://doi.org/10.32473/edis-hs291-2005

[44] Ramos, F., Osorio, C., Duque, C., Cordero, Aristizábal, F., Garzón, C. & Y. Fujimoto. Fujimoto (2004). Estudio químico de la nuez del

marañón gigante (Anacardium giganteum*). Revista Académica de Colombia y Ciencia*, 28 (109), 565-575.

[45] Sindoni, M, Marcano, L., & Parra, R. (2008). Studies on the acceptance of merey-derived harins for the preparation of breads. Agronomía Tropical, 58(1), 11-16. Available at: http://ve.scielo.org/scielo.php?pid=S0002-192X2008000100003&script=sci_abstract

[46] Pérez Morales, M.L. y Velázquez. F. (2021). Artisanal production of citric acid and ascorbic acid from caujil (Anacardium occidentale L.) and mango (Magnifera indica). (Special Degree Work for Chemical Engineers). Rafael Urdaneta University. Venezuela.

[47] Sindoni, V. M. J., Caldera, R. E., Pérez, A. C., Marcano, L., Parra, R., & Marín, R. C. (2007). Evaluation of coagulating agents for the formulation of juice from merey pseudofruits. Agronomía Tropical, 57, 61-65. Available at: http://ve.scielo.org/scielo.php?script=sci_arttext&pid=S0002-192X2007000100008

[48] Tuler, A. C., Peixoto, A. L. & Barboza da Silva, N. C. (2019). Unconventional food plants in the rural community of São José da Figueira, Durandé, Minas Gerais, Brazil. Unconventional food plants in the rural (UFP) community of São José da Figueira,Durandé, Minas Gerais, Brazil. Rodriguésia, 70, e01142018, 1-12. https://doi.org/10.1590/2175-7860201970077

[49] Brito, E. S., Silva, E. O. & Rodrigues, S. (2018). Cashew - Anacardium occidentale. In S. Rodrigues, E. O. Silva & E. S. Brito (Eds.). Exotic fruits: reference guide. Academic Press.

[50] Sindoni, M., Hidalgo, P., Chauran, O., Chirinos, J., Bertorelli, M., & Salcedo, F. (2005). Merey cultivation in eastern Venezuela. Series manuales de cultivo INIA. National Institute of Agricultural Research.

[51] Murga-Orrillo, H., Coronado Jorge, M. F., Abanto-Rodríguez, C., & Lobo, F. D. A. (2021). Altitudinal gradient and its influence on the edaphoclimatic characteristics of tropical forests. *Madera y Bosques*, 27(3), e2732271, 1-13. https://doi.org/10.21829/myb.2021.2732271

[52] Borjas, M. (2015). Semi-industrial-scale production of a jam from the pseudofruit of the caujil *(Anacardium occidentale L.)*. (Master's Thesis in Food Science and Technology). University of Zulia. Venezuela.

Printed by Books on Demand GmbH, Norderstedt / Germany